A Creek, a Hill, & a Forty

The Early Years of Alaska's Matanuska Colony,
seen through a colonist's letters home
—
Margaret Miller's story

Ray Bonnell

Pingo Press
2024

Copyright 2024 by Ray Bonnell.

All rights reserved. No part of this publication may be reproduced; stored in a retrieved system; transmitted in any form or by any means—electronic, mechanical, photocopying, or otherwise; or used in any manner for the purposes of training artificial intelligence technologies to generate text, without the prior written permission of the publisher.

Book design, layout and formatting done at Pingo Press. Fonts used: Alexa, Gil Sans, Minion Pro

Cover photo: The Miller house soon after it was completed in 1935

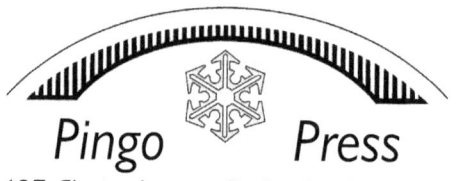

127 Glacier Avenue, Fairbanks, AK 99701
info@pingopress.us

Printed in the United States of America
First Printing, August 2024
ISBN 978-1-7364236-6-0

Miller Family in 1940 (L-R) Margaret, Priscilla, Neil, Tim. Mardie, Janell

Contents

Preface	vi
Prologue	1
1 Surely you aren't going to that God-forsaken Country	13
2 Alaska Bound	24
3 First Impressions	36
4 Sensational Reading	53
5 Settling In	68
6 Under Roof	88
7 This is Fun	104
8 Matanuska Valley Christmas	119
9 Mid-Winter Progress	131
10 A Marvelous Winter	142
11 Springtime in Alaska	154
12 Birthday Celebration	166
13 Triumphs and Tragedies: 1936-1937	180
14 The Rest of the Story: 1938 Onward	196
Appendices	214
Graphic Image Credits	242
Notes	245
Bibliography	255

Preface

A Creek, a Hill, & a Forty provides a glimpse of the early years of Alaska's Matanuska Colony, but it is not intended to be a comprehensive history. If you are looking for general histories of the project you should read books such as Evangeline Atwood's *We shall be Remembered*, or Helen Hegener's *The 1935 Matanuska Colony Project*. For a more in-depth analysis of the colony's history, O. W. Miller's *The Frontier in Alaska and the Matanuska Colony* is a good choice.

This book, rather than just being a synthesis of historical information garnered from books, government reports, newspaper and magazine articles; is also the personal narrative of one colonist family, told primarily through the letters of Margaret Miller, a colonist wife.

When planning for the Matanuska Colony began, Margaret was a housewife living in Wisconsin with her husband, Neil, and their three daughters. But she was far from a typical housewife. Margaret, trained as a teacher, was an erudite and accomplished writer. Beginning with the family's move to Alaska in 1935, and for several years after that, she wrote multi-page letters to her mother in River Falls, Wisconsin (sent stateside on the weekly Alaska Steamship Company steamer leaving Seward headed for Seattle). Those letters, in addition to the progress of the colony, reveal much about everyday life in the Matanuska Valley.

Many of Margaret's letters were published in the *River Falls Journal*. Her mother kept all of those letters, as well as clippings related to the Matanuska Colony that appeared in newspapers and magazines, and she eventually returned the letters and clippings to her daughter. The family still has Margaret's letters, and they form the core of this book—a book that Margaret wanted to write, but never had the resources or time to finish.

I picked up the mantle of writing Margaret's book after marrying her granddaughter, Elizabeth Bacon. Knitting together

Preface

many of Margaret's letters, along with some of her later writings, I have hopefully created a readable narrative of the Millers' lives in the Matanuska Valley. Occasionally I added in details that Margaret may not have been aware of, or to explain gaps in the narrative or 1930s' references that might seen obscure to modern readers. Any opinions expressed in Margaret's letters are solely her own.

Much of the information that Margaret passed along about matters outside her family was apparently gleaned through her family's friendships with colony administrators; by reading the two small newspapers published at Palmer, the *Matanuska Valley Pioneer* (1935-1936), and the *Valley Settler* (1937-1950s); and by being involved in much of the social life of the valley. I have found Margaret's letters to be a reliable reflection of life in Palmer, and wherever possible, I have corroborated her statements through outside sources such as project documents and articles in the *Valley Pioneer* and *Valley Settler*.

This book would not have been possible without the support of Margaret's family and complete and continued access to Margaret's papers, which include the letters themselves, later writings of Margaret, copious amounts of newspaper clippings, calendars, the minutes from early colony council meetings, miscellaneous correspondence, and photographs.

I am indebted to my wife, Betsy, for her services during the book's development. She is a skilled editor, and guided the book from developmental edits to final proofreading.

I want to acknowledge the contribution of Joannie Goeser, of Goeser Graphics, for providing the maps included in the book; and would also like to acknowledge the role of Palmer historian, Jim Fox, who transcribed many of Margaret's letters before I became involved with the project.

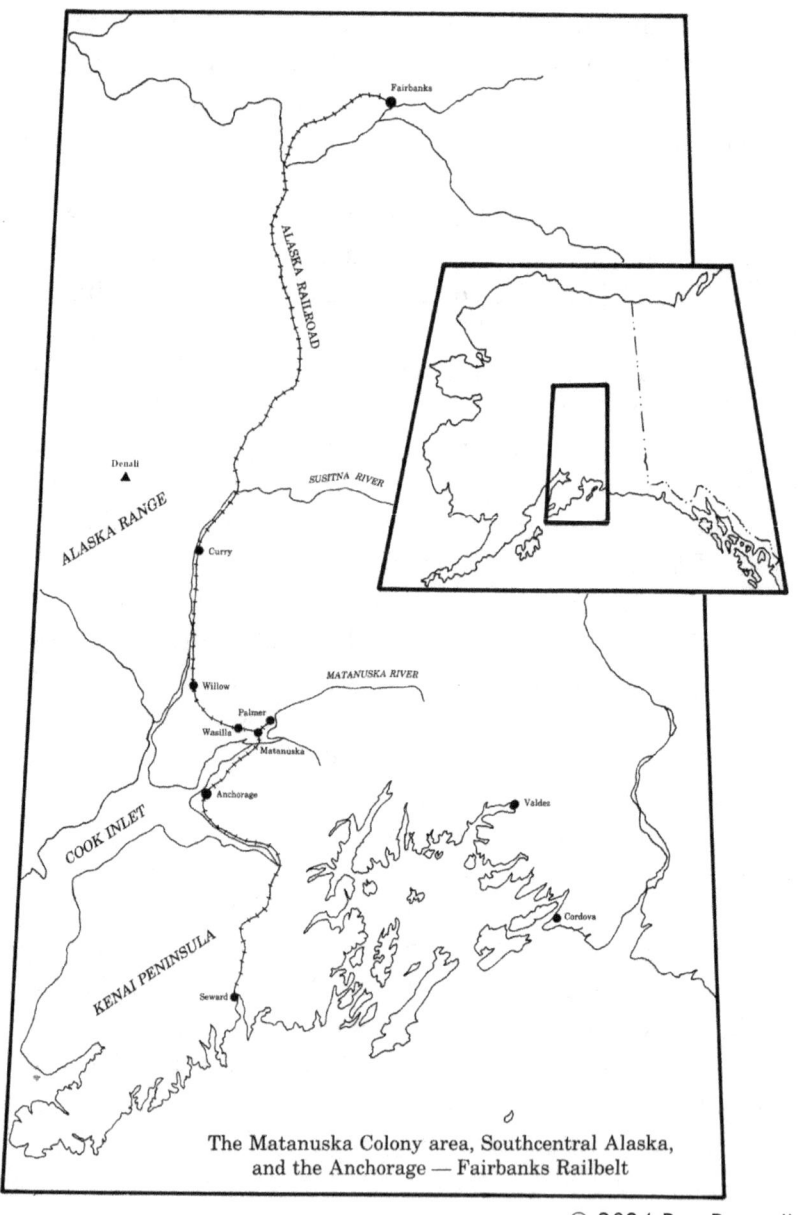

The Matanuska Colony area, Southcentral Alaska, and the Anchorage — Fairbanks Railbelt

© 2024 Ray Bonnell

Prologue

History of Matanuska Valley and Matanuska Colony prior to arrival of Michigan and Wisconsin contingents

When the Federal Emergency Relief Administration (FERA), one of President Franklin Roosevelt's New Deal programs, announced in the spring of 1935 that it would move 200 farm families from the Great Lakes area to the Matanuska Valley in Alaska, the first question asked by many people across the nation was, why?

It wasn't a question of why move the families. For several years, the federal government, through FERA and other agencies, had been relocating some of the nation's poor in an effort to help them rise out of poverty. Under a few FERA rural resettlement projects, such as the one in the Matanuska Valley, new farms were established and small agricultural communities were built to provide the farmers with needed services. While the Matanuska Colony, as it came to be called, was one of the largest such projects, it was only one of about 100 resettlement projects throughout the nation.

Resettlement was only a minor part of the New Deal's program to assist and rehabilitate the country's poor, but the resettlement of families to the Matanuska Valley in Alaska received an inordinate amount of national interest and considerable play in the nation's press.

What people wanted to know about the Matanuska Colony project was why move 1,000 people all the way to Alaska when there was plenty of empty land elsewhere in the country to resettle.

Their question was coupled with the federal government's estimate that it would cost approximately $1,000,000 ($22,000,000 in 2024 dollars) to resettle the families, develop their farms, and build a new community to serve their needs. (The total cost for the project would eventually exceed $5,000,000.)[1]

Some of the questions arose because of the general lack of knowledge about Alaska (a still not uncommon problem). For instance, although the federal government had operated agricultural experiment stations at numerous locations throughout the territory for decades and knew the territory's agricultural potential, some folks, without any evidence other than hearsay, argued that crops couldn't be grown in Alaska.

The reasons behind the actual selection of Alaska for a farming colony were many. Territorial Governor Ernest Gruening, in his 1954 State of Alaska speech, mentioned that the colony was established to give farmers in stricken agricultural areas a new chance, to stimulate the growth of Alaska's population, and demonstrate the agricultural potential of Alaska. Others pundits had reasoned the colony would provide needed food and other supplies for the military developments planned for the territory.

Another inducement for the federal government to move farmers to Alaska was that the area was so far from the nation's major agricultural markets that any goods produced for sale in the valley could not have any impact on the existing overabundance of agricultural goods in the rest of the nation.[2]

Historian Orlando Miller writes that given the background of the situation "...the back-to-the land movement that was widely publicized during the early depression years, the realization of the chronic nature of much rural poverty, the existing New Deal resettlement and homestead programs, the thirty years of advertising Alaska's undeveloped potential for settlement, and most important, the Alaska Railroad program for encouraging settlement...it would have been surprising had Alaska been overlooked as a possible site for resettlement."[3]

The reasons for choosing Alaska for a new resettlement farming community have roots reaching back to at least 1914, a date which coincides with two events greatly affecting Alaska:

Prologue

the beginning of World War I in Europe, and the authorization of the Alaska Railroad by Congress.

From the war's outbreak, Alaska was strongly pro-Allied, and even before the United States officially joined the war, many young Alaskans joined the war effort: men joining Canadian units to fight in Europe, and women joining the Red Cross. After the United States entered the war in 1917, even more of the territory's residents answered the call to serve. Others moved south to work in highly paid war-related industries.

The sudden lack of workers forced labor prices up, and the increased cost of mining machinery and supplies due to the war caused many marginal mining operations to fold, further exacerbating Alaska's already-slumping economy.

The end of the war did nothing to increase Alaska's population or alleviate its economic problems. Many of the territory's former residents stayed in the Lower 48 States to work in the post-war boom economy. Furthermore, Alaska's problems were compounded after the war by the drop in demand for two of the territory's chief exports, copper and salmon.

Between 1910 and 1920, the territory's population plummeted by almost 13 percent, from 64,356 to 55,036.[4] Many Alaskan businesses, local government leaders, territorial officials and some federal officials felt that this dearth of population was reason enough for the Matanuska project to proceed.

Another factor affecting FERA's decision to establish the Matanuska colony was the Alaska Railroad, constructed between 1914 and 1923. The railroad was an amalgamation of an earlier railway line running from Seward to the Turnagain Arm of Cook Inlet, and new lines running northward from the end-of-tracks (near present-day Girdwood) through Anchorage and on to Fairbanks in Interior Alaska.

It was constructed to open Alaska to development, especially development of the coal deposits located along the railroad route. Railroad construction brought a temporary increase in the railbelt's population as well as a temporary lift to the area's economy. However, with only a few exceptions (primarily related to mining), until the mid-1930s it did little to expand Alaska's economy or increase its population.

Ship Creek, located across Cook Inlet from the Matanuska Valley, was established as the railroad's field headquarters in 1914. Hopeful railroad workers and others seeking new opportunities quickly flooded the area, and the Alaska Engineering Commission, which was responsible for construction of the railroad, laid out the Anchorage townsite the next year.

It was soon a bustling city by Alaska standards and by 1916, during the height of railroad construction, Anchorage had an estimated population of over 6,000 people. In 1920, with railroad construction winding down, Anchorage's population had shrunk to under 2,000.[5]

During the construction phase, and as the completed tracks worked their way northward, there was a small influx of settlers to the territory, but by 1920, only 364 farms and 5,736 improved acres had been developed along the entire railroad. A decade later, Alaska still had not recovered from the loss of population after the outbreak of World War I. In 1930 the populations of the entire railbelt, extending 470 miles from Seward to Fairbanks, was only about 6,400 people.[6]

The Alaska Railroad regularly ran at a loss, and its officials knew that it would never be successful unless the land adjacent to the railroad right-of-way was developed. However, nothing was done for many years to stimulate growth in the area. When Colonel Otto Ohlson became general manager of the Alaska Railroad in 1928, he initiated a program to promote the long-desired development of the area.

Little development had occurred in the Matanuska Valley prior to construction of the railroad. Trappers and traders had ventured into the Cook Inlet area since the Russian Period, but they fostered few permanent changes.

Prologue

A small gold stampede along Cook Inlet that began in about 1896 ushered in the modern development of the upper Cook Inlet. Prospectors fanned out from the initial gold strikes on the Kenai Peninsula into the Knik Arm region, and one area they explored was the southern flanks of the Talkeetna Mountains.

The first placer-gold claims in the Talkeetnas were staked along Willow Creek in 1897, and lode-gold veins were discovered nearby in 1906. While the placer claims were quickly exhausted, hard-rock mining continued in the area until the 1950s.[7]

The town of Knik on the northwestern shore of Cook Inlet's Knik Arm, which began as a Dena'ina Athabascan village, grew into the commercial and transportation center for the area, primarily to support the Talkeetna Mountains mines, but also as a transportation link along the Iditarod Trail. By the mid-1910s Knik had a population of about 250 people and the Matanuska Valley had about 400 residents.[8]

With the assurance that the Alaska Railroad would be built through the valley, people rushed to stake homesteads along the railroad, even as the right-of-way was surveyed and brushed out. This prompted the federal government to establish an agricultural experiment station nearby. The Matanuska Experiment Station, located about a ½ mile from the railroad and only a few miles from the settlement at Matanuska, was approved in 1915, but work on the station did not start until 1917.[9]

As track-laying crews worked their way northward, railroad townsites were established at Matanuska in 1915, and at Wasilla in 1917. Wasilla was located where the main wagon road from Knik to the gold mines in the Talkeetnas crossed the railroad, and the valley's center of activity and population quickly shifted from Knik to Wasilla. By 1921, most of the agricultural land in the valley that was easily-accessible (generally, land adjacent to waterbodies, to the railroad, and to roads to the mines) was homesteaded.[10]

ARRC Staff - May 1935
(Top row, L-R) J. Givens, Asst. Property Custodian; H. Lying, Asst. Manager; Don Irwin, General Manager; E. Cronin, Dispersing Officer; Stuart Campbell, Property Custodian; Ross Sheely, Construction Superintendent.
(Bottom row, L-R) R. Atwood, Chief Clerk; R. Hare, Asst. Accountant; O. Cowden, Commissary Manager; D. Sullivan, Accountant; Eugene Carr, Procurement Manager; A. Betts, Chief Assistant; Francis Biggs, Engineer.

Under Olson's leadership, the railroad had some success in attracting settlers to Alaska. Between 1928 and 1934, 110 settlers started homesteads in the Matanuska Valley. (For a short explanation of the federal homesteading program see endnote [11]) Homesteading was difficult, though, and many of the farms were little more than gardens. Some settlers gave up quickly. Others left after receiving title to their homesteads, returning to paying jobs elsewhere.

Alaska's lenient homestead laws and the absence of land taxes even encouraged settlers to leave. Once they had title to their land, settlers could quit their homesteads, as well as leave

Prologue

Alaska, without fear of losing their land. Of the over 550 homesteads started in the Matanuska Valley between 1914 and 1934, only about 117 were occupied in early 1935.[12]

Alaska's population slowly began increasing as the nation's economy stalled in the early 1930s and unemployment in the Lower 48 States increased. Former residents began returning to the Territory, accompanied by some newcomers, but the territory's population still fell below its pre-World-War-I level.

By the mid-1930s the dream of a new life in Alaska still appealed to many people throughout the United States, but the number of new settlers diminished as the nation's economy worsened. Prospective settlers could no longer afford the expensive move. The Railroad continued to send out pamphlets advertising settlement in Alaska but stopped its aggressive settlement campaign.

Although private development of Alaska may have lagged in the 1930s, the federal government, perhaps spurred by all those development pamphlets produced by the Alaska Railroad, began showing interest during the 1930s in developing the region. The military, aware that Alaska was on the great circle route to Asia, and noting its possible strategic importance, seemed especially interested.

Other branches of the federal government were also becoming interested in Alaska. Jacob Baker, assistant administrator for FERA, toured Alaska in summer 1934, and included a visit to the Matanuska Valley. He came away enthusiastic about the possibility of agricultural development there and met with Colonel Ohlson, chairman of the Alaska Railroad, and Territorial Governor John Troy, on his trip back to Washington, D.C. All three were optimistic about establishing a farming community in the valley.

The Matanuska Valley was chosen over other Alaska sites as the location of the colony partly because it was close to Anchorage, a possible market for its goods. There was also an agricultural experiment station with 20 years of experience in

Matanuska Experiment Station—1930s.

valley agriculture located near the colony site, and the area had a relatively mild climate.

Proximity to the Alaska Railroad was also an important factor in siting the colony, since any potential project site had to be easily accessible. The town of Seward had lobbied for an agricultural project on the Kenai Peninsula. However, even though the railroad ran across the peninsula, its tracks were miles away from any potential agricultural area and there was no road access.

Accessibility was not a major problem in the Matanuska Valley. The railroad passed through the southwestern edge of the 60-mile-long Matanuska Valley on its way north, and a spur, which ran northeast through the valley's center to coal deposits in the Talkeetna Mountains, branched off from the main line at the tiny settlement of Matanuska.

When Baker returned to Washington, a flurry of activity related to a new agricultural community for Alaska ensued. Events developed rapidly and by January 1935, the Matanuska Colony had been approved by Washington, D.C. Don Irwin, the

Prologue

Another view of the Experiment Station.

supervisor of the federal government's Matanuska Experiment Station, was called to Washington, D.C. in early March to assist with planning for the project, and he was appointed the project's general manager.

The Alaska Rural Rehabilitation Corporation (ARRC) was established to manage the project's development. The ARRC, usually referred to as the Corporation, was incorporated as a non-profit corporation on April 12, 1935. It was organized along the lines of corporations already set up to manage other New Deal resettlement communities.[13]

Because of the similar climates, state relief agencies (at the behest of the federal government) selected colonists for the Matanuska Colony mainly from the relief roles in the "cut-over region," as the northern counties of Wisconsin, Minnesota, and Michigan were called. The economy of the entire nation was suffering in 1935, but the cut-over region was in poor economic condition even before the Depression.

By 1935, the cut-over region was one of the most severely depressed areas in the nation. The number of residents on relief varied, but some counties had up to 70 percent of the population on relief. In Wisconsin, the average income from farming in the northern counties was half that for the rest of the state. By June 1935 when the national relief rate was 8.8 percent, the rate for the entire cut-over region was 18.4 percent.[14]

The New Deal's programs had little effect on the stagnant economy of the area, but the Matanuska Colony did offer escape to a few of the unfortunate in the area. According to the initial plans for the colony, all colonists were to be chosen from the relief rolls, and preference would be given to people of northern European stock because of their assumed hardiness and pioneering spirit. All colonists were also to have a cooperative nature. Families were to be of average size, preferably with three to four children. Adults to be between 25 and 35 years of age, with some farming experience, but hopefully also with specialized skills that would be useful in a small agricultural community.[15]

But many of the families chosen had little farming experience. Some families on relief had returned to rural areas because of the unemployment in the cities, moving in with relatives or squatting on vacant land. Some long-time inhabitants worked regular jobs more than they farmed. Because of their desire to qualify for the resettlement program, many applicants embellished their applications to expand a few years of gardening into substantial farming experience.

There were 201 families (903 people) chosen from the lake states (68 from Wisconsin, 67 from Minnesota, and 67 from Michigan). The people selected to participate in this grand adventure didn't know the reasons behind its location and probably didn't care. They just wanted to go.

In preparation for the colony, President Roosevelt, in Executive Order No. 6957, withdrew 11 townships in the valley on February 23, 1935 as a colony preserve. Out of those townships, 241,332 acres were actually affected by the withdrawal. The rest of the land, some 12,000 acres, was already in private ownership or was under application for homesteads or mineral entry.[16]

Prologue

FERA officials wanted the colony established as quickly as possible. To speed the process they sought to buy land already cleared and in production. They sent Ross Sheely, director of the University of Alaska Extension Service, to the valley in April 1935 to select and survey tracts for the colony, and to acquire options on as much improved land as possible. He purchased about 7,500 acres, which included both cleared and uncleared land.

Of the cleared land purchased, most was near the nascent community of Wharton, which in the early 1930s consisted of a few homesteads, a railroad siding, and a small freight warehouse next to the tracks. Wharton (name later changed to Palmer), was chosen as the base of operations for the project. It was there where the main colony facilities would be constructed, and where, for the summer of 1935, the main camp for housing colonists would be located. After setting up camps and allowing space for community gardens, only 175 acres of cleared land for other crops remained when the colonists arrived.[17]

Over 400 "transient workers," the temporary workers (selected from relief rolls) who preceded the colonists to Alaska to build project facilities, arrived in Alaska in April and early May.[18] Their initial job was to prepare camps for the arrival of the colonists. Nine camps were established: the main camp along the railroad tracks at Palmer, and eight outlying camps. Work included erecting over 200 sixteen-foot by twenty-four-foot canvas tents in which the colonist families would live until houses were built.

After the arrival of the colonists, the transients would build administrative and other colony facilities, cabins and barns, drill wells, and clear fields, working alongside the colonists to get the colony in shape before winter. Most of the transient workers would then return to the States.

Some Alaskans were outraged that workers were imported for the project. However, project officials responded that there weren't enough workers to be found in Alaska, the type of workers needed were not available, and any workers already in Alaska or migrating there would probably be absorbed by the mining and fishing industries. Alaska officials did anticipate an influx of workers lured by the project, and Territorial Governor Troy warned job seekers to stay away.[19]

The West Coast profited considerably from the Matanuska Colony. Seattle, Washington, which had been the main port for goods shipped to Alaska, expected to be the recipient of much of the business, but San Francisco appeared to be the early winner in the West Coast sweepstakes to supply the colony. The first shipload of transient workers was chosen from California's relief roles and the transport ships to take the people and supplies to Alaska were chartered out of San Francisco.

The Bureau of Indian Affairs supply ship North Star and the U.S. Army troop carrier St Mihiel were the two government ships chartered for the project. The North Star left San Francisco April 23 with 1,315 tons of cargo, 118 transient workers, and supervisory and support personnel. It arrived at Seward May 6. The Minnesota contingent of colonists left San Francisco on the St Mihiel on May 1.[20]

FERA justified its decisions for basing much of its early operations out of California rather than Washington in several ways. First, it chose workers from California because the relief roles in California were so much larger than in Washington. Second, the transport ships were chartered out of San Francisco because bids from commercial shipping companies were too high, and the government ships available were based in San Francisco.

Elected officials and businessmen from Seattle and the rest of the state of Washington reacted immediately. Alfred Lundin, Seattle Chamber of Commerce president, said using workers from California and shipping from San Francisco were uneconomical and ridiculous. The protests evidently had some results. The next load of men and supplies were shipped from Seattle, and the Wisconsin and Michigan colonist contingents left from there on May 18.[21]

One

Surely you aren't Going to that God-Forsaken Country?

The St Mihiel wallowed pregnantly through the heavy seas of the Gulf of Alaska as I pushed the girls before me and stepped from the crowded, stifling lounge out onto the rolling deck.

Hunching my coat collar up, we began another "stroll" around the deck. Below deck we were oppressed by the humid air—stuffed with the odors of vomit, dirty diapers, and greasy food. When we ventured on deck the spindrift pierced our clothes, and stung our faces and hands, but at least the air was clean. All in all, our voyage to Alaska was absolutely horrendous.

There were 605 of us, including my husband Neil, our three daughters; Mardie, Priscilla, and Janell, and myself, all crowded aboard the U.S. Army transport St Mihiel. We were "colonists," bound for Alaska as participants in the Matanuska Colony, a New Deal resettlement project. Many of us had never been on a ship, and we were very apprehensive about the trip. Neil and I had been told that the ship would follow the scenic Inside Passage to Alaska, and had hoped the ocean voyage would be an enjoyable sightseeing trip.

But how enjoyable could it be for anyone on a ship overloaded with 3,800 tons of freight, and passengers crammed into every cabin and spare hold. The men and women were segregated into separate hatches, below the water line.

To get to our hatch we had to walk past the kitchen, down a ladder, past the bathrooms, and down another steep stairway. The only ventilation was forced through ducts that I swear brought air from the kitchen and bathrooms. Mothers with small babies were a little better off—they were crowded into staterooms above the water line.

Then, instead of cruising the Inside Passage, we cut across the Gulf of Alaska, straight into a spring storm with 12-foot swells. Most of the passengers, and even some of the crew became sick, and sanitary conditions aboard the ship quickly deteriorated.

Neil kept walking us back and forth along the deck to prevent seasickness. One little girl on board watched us curiously and asked, "Why do you walk all the time? You don't have to walk to Alaska." But the walking did help, and we never got seriously ill, although we were still so miserable. We just kept looking forward to our new life in Alaska, and that somehow got us through.

The ocean voyage certainly forced us to at least temporarily question why we were "immigrating" to Alaska. Our curiosity about the project was first piqued in March of 1935. I came home from Sunday School one morning and found Neil going through our atlas and encyclopedia. As I entered he asked me how my pioneer blood was. As usual, I answered that oft asked question very eagerly, for I knew we were off to another land of imagination, where we were pioneers starting from scratch and building a life in some newly settled area.

My thoughts quickly bounced back to reality when Neil showed me an article he had found in *The Milwaukee Journal* about the United States sending a group of 200 families to the Matanuska Valley in Alaska.

Because of the similar climates, colonists for the Alaska project were to be drawn mainly from the relief roles in the "cut-over-region," as the northern counties of Wisconsin, Minnesota, and Michigan were called.

According to the newspaper article, the people chosen for the project were to be of northern European stock because of their supposed ruggedness and pioneering spirit. Families were to be fairly young and of "average" size. Adults were to be between 25 and 35 years old. Men would be chosen primarily for their farming experience, but consideration was also given to other valuable skills such as carpentry, plumbing, and fishing. All families were to be chosen from the relief roles.

Each family was to be given forty acres of land (three of which were to be cleared) two cows, a few chickens, and be helped financially for the first year. The total cost per family would range around $3,000, to be paid back in 30 years. There would be no payments on the interest or principal the first five years. In addition to developing these farms, FERA would build a new community to serve all the colonists' needs. These people were to be given a chance to start anew.

Some way it seemed to us that if we could only be a part of the project, we could get our start to live as we had always wanted to live. Even before we were married, we planned to "someday" have a "40" and live the pioneer life. Someway the thought of starting at the very bottom and building up everything, through our own efforts and dreams, had always seemed ideal. To be able to look around us and say this is ours, we built it, and it is our plan. To be able to say what we have, we have because our working together has brought this into being and made it what it is. To be able to plant trees and flowers and create a pleasing surrounding that reflects our own personalities. To feel secure—to not have to live from the beginning of the school year to the end of the school year, wondering where Neil would teach and where we'd be the next school year. Those things meant much to us.

We were then living in Blair, Wisconsin, just south of the cut-over region. Neil was the principal of the city's high school and I was a housewife, although I had training as a teacher.

We and our daughters lived a fairly comfortable life, at least when compared with the residents of the depressed cut-over-region. But Neil's life as a schoolteacher in Wisconsin then was far from secure. Because of the high relief rates and stagnant economy, few property owners were able to pay the taxes which supported the schools.

Teachers were not well paid, and Neil refused to dabble in the local politics, so we moved often to new teaching jobs,

A Creek, a Hill, & a Forty

The Millers before leaving Wisconsin for Alaska.
(L-R) Janell, Neil, Mardie, Priscilla, Margaret.

spending only two or three years in one place. No sooner would I fix one house up than we would move. At one point Neil even gave up teaching and returned to work on his parents' farm. We often dreamed of settling somewhere with a creek, a hill, and a forty.

When the opportunity to go to Alaska came, we could not pass it up. We knew there was little chance of being chosen to go. Neil was not on relief and we did not live in the northern counties, but we had to try. The state relief agency tried to discourage us from applying, but we filled out the forms anyway. Although we probably should have been discouraged, Neil or I always found some hope in little phrases in the letters from the relief agencies. After all, Neil did have farming experience, and teachers would be needed in the colony. Because there was little chance of being chosen, we dared not hope for a miracle. But our optimism still remained.

We finally received a telephone call from the relief agency May 1. I could hardly believe our luck. It had taken a special dispensation from Washington, D.C., but our application had been approved. We had actually been chosen to go to Alaska!

Surely you aren't Going to that God-Forsaken Country?

I had such conflicting emotions! Feelings of elation, glory, thanking of God—all mixed with little pangs of something. Regret? Hardly regret, for we were too thrilled and anxious to go. But I knew that somewhere tears would come.

When the telephone call came, we had a hard time to decide just what our emotions were trying to do. We were so happy. So of course I cried. But I was a little surprised upon going into the kitchen a few minutes later to find 13 year old Mardie laughing through her tears.

The girls were so thrilled over the plans. We asked them not to tell it at school just yet. When Janell (nearly 9 years old) came home that next night she said, "Gladys knew already. She told Miss Larson." So I knew Janell had not been able to "keep" it. When Priscilla (11 years old) came home, we asked her if she had managed to keep the secret, and she looked sort of guilty and said "Pearl guessed." Bless their hearts! I could not keep it either! Right away, I had to telephone my friend Ethel and tell her. I felt I just had to, so I just know how the girls felt.

The agency provided no time for lengthy celebration. We were to depart May 14 along with the other 134 families chosen from Wisconsin and Michigan. The 67 families from Minnesota had already been selected and were on their way to Alaska! Everyone had precious little time to pack and say farewells.

Two weeks to prepare for a 4,000 mile trip to a region we knew little about aside from the fact that the climate was similar to that of Wisconsin! After the agency's phone call, Neil and I scoured the school library for information about Alaska, but we found little aside from a map of the territory showing the location of Cook Inlet and the Matanuska Valley, where our new farm would be.

The state relief agencies and the Federal Emergency Relief Administration (FERA) did fill us in about what we could expect in Alaska, and what the project would be like, but the public was generally misinformed about life on the Alaskan frontier. I received a letter from a friend saying, "As I was reading an *Eau*

Claire Leader, I happened to see an item where a number of families were going to Alaska, and reading a little further I see where a Neil Miller was one to go. It surely isn't your Neil that expects to go up there in that God-forsaken country. Surely you and the girls aren't going! Saying it just makes me sick. I just can't imagine you folks going up there to live with all those Finlanders. When I read it I could have cried."

Perhaps our hardest job in preparing for the trip was parting with cherished family furnishings and mementos. The 2,000 pound limit on each family's personal goods shipped to Alaska meant that most of our furnishings and possessions had to be left behind, in preference for necessities such as bedding,

Neil's tools, my sewing machine, clothing, cooking utensils, and toiletries. Every once in a while we would realize some new little pang of regret sneaking in, over having to leave something. We simply could not take Grandma's clock, and we felt it was almost part of us! And our living room set, that I liked so well, could not go. They told us to bring springs but not mattresses, and so it went. I was glad our time was short so we had to hurry, or some of those regrets might have had time to breed.

While we left most of our possessions behind, we could not part with the piano. Into the piano and its packing crate we stuffed bedding and extra clothes. Packing items inside the piano seemed a good idea at the time, but for years after that a strange note or "dead" key would find us pulling socks and other clothes out of the piano.

The job of preparing for the trip probably would have been even worse if not for the friends and family who helped pack, provide meals, and care for children. With the packing, disposal of excess furnishings, farewells to friends, and tying up of loose ends, everything was often in chaos.

Every few minutes someone came or telephoned or sent a farewell gift, and I did not seem to accomplish much of anything. Local friends coming and going all the time added to the confusion of it all, but it was so good to have them.

Surely you aren't Going to that God-Forsaken Country?

One night we were too tired to relax to go to sleep, and finally got up and took hot baths hoping thereby to quiet our bodies into relaxation. We were just settling down again, all ready for a good sleep, when a car honked outside. There was Neil's army pal Alex, and his wife and some friends from Eau Claire. They had come down to give us a sendoff. So we did not get to bed again until two o'clock. But oh, it was good to have friends there! I think the fun and comfort of their coming did us more good than sleep would have.

After all our belongings were either packed, given to friends and relatives, sold, or otherwise disposed of, we had to get from our homes to the West Coast, where a steamer was waiting to take us to Alaska. We were to meet the other members of the Wisconsin contingent in St. Paul, Minnesota on May 14.

The nearest railroad station to Blair was 30 miles away in Winona, Minnesota, and we stayed at a hotel there the 13th, in order to catch the morning train to St. Paul. The girls started sleeping three in a bed, but Priscilla climbed in with Neil and me before morning.

It was such a good comfortable happy feeling to be all together, on our way to the sort of life we had always wanted. Neil counted our money that night. We started to Alaska with $53.55.

The St. Paul train station was certainly a lively place when we arrived the next morning. The platform was crowded with departing families, friends and relatives saying good bye, and reporters and photographers pushing in for stories and pictures.

We colonists could easily be picked out by the colorful ribbons on our lapels. Everywhere we looked we saw families with ribbons on their coats. I wondered if the rest were sizing me up—as I was them—wondering then who would be my neighbors and how I would like them—wondering what attracted them and what their background was.

All the people seemed interesting to me. They were all so eager to be "at it." They were not necessarily farmers, however, as we supposed they would be. They seemed to be from all walks of life. Some were not even much familiar with farms—but all seemed willing and anxious to try.

Marjorie Case, a social worker who accompanied the Michigan contingent to Seattle, confirmed my suppositions in a *Seattle Post-Intelligencer* article. Case wrote that the colonists, "... all have farm backgrounds, but they are not all farmers. There are carpenters and trappers and woodsmen. They are going to be a self-sufficient little community among themselves.

"They were not all stone broke. Some were what we call borderline cases. They were chosen for their physical and educational qualifications, their character and their social adaptability."[1]

My mother and family friends were at St. Paul to see us off. It was not so bad at the station, but later that night, after everyone else on the train was asleep, there was too much time to dwell on goodbyes. Having to say goodbye to so many prolonged the ceremony too long. (But it was swell their being there, and by walking straight away without looking back, I managed to control it, until I was around the corner and out of sight.)

We rode in day coaches during the two day trip to Seattle. The cars did not even have reclining chairs, and were they dirty. Some families, with tiny babies too, had been on the train since the morning before at 11 o'clock. The poor kids were so tired and cross and dirty—but there was not one, but who told you with a gleam of joy in their eyes that "We are going to Alaska!" One little fellow ate part of an ice cream cone and refused to eat the rest because he wanted to save it until he got to Alaska.

Later that night many of the men discovered how to take the seats apart to improvise beds. We, like most of the people, lacked blankets or pillows and simply slept in our clothes.

Our train had 17 cars, all of them crowded. Twenty-one people occupied our car throughout the two day trip to Seattle. The train stopped occasionally for provisions, to let mothers wash diapers, and let people off to stretch their legs, but otherwise, we all ate, slept, played, and whiled away the time on the train. Unfortunately, several children on the train came down with measles and had to be isolated.

Our group was accompanied by a Mrs. Gamble from Washington, D.C. who was supposed to take care of us. She walked up and down the aisles, usually accompanied by Arville Schalaben of the *Milwaukee Journal*, inquiring about each family's health and needs.

Gamble did not really have any nursing or social work experience, and some people told me she was really a sort of "spy" sent by Washington to see what conditions were like on the trip, as the Minnesota contingent was already sending complaints about the project back to Washington. But she did minister to our needs and we appreciated her help. One of the babies on the train became seriously ill and Gamble had the train stopped so the child could receive proper medical care. That shows you how well we were cared for.

Feeding everyone was a major problem because of the crowded conditions on the train, and we certainly had crazy hours for meals. We did not have lunch until 1:30, supper at seven, and some people had been on the train since five in the morning without food—even babies.

But the service and staff were commendable. The diners were always clean and the service lovely. The waiters were very thoughtful of us and made us forget for the time that we were "emigrants" and made us feel as though maybe we rated after all. I did not hear anyone complain about the service.

The train trip was uncomfortable for most of us, with inadequate washing facilities, crowded cars, and little to do. Priscilla told me she counted 59 train tunnels between St. Paul and Seattle. We did have plenty of time to become acquainted with each other, though.

The popular form of light amusement was wondering who our neighbors would be. I did not meet any people in our car who weren't interesting and congenial. The Bailey, Barry, Campbell, Ring, Nelson, Anderson, and La Flamm families were in our car. They quickly became familiar faces although we couldn't remember all their first names.

Watching the children on board play and get to know each other provided some of the most enjoyable moments on the trip. There was one little girl, three year old Iris Ring, with round face dirty from travel, black eyes dancing with excitement, who watched out of the train window as the train started out of one station and said, "Oh, oh. Look at all the houses and barns going. Going to Alaska!"

The trip also provided a few rare moments of solitude. On the second night out, Neil and I woke up almost suffocated. All ventilators were closed and steam was on full blast—and 21 people breathing into the air. Neil and I got up and opened the back door and fixed the ventilators—then went out on the back platform for air.

We were going through Idaho in the Coeur d'Alene country and were fortunate enough to see the moon on the water of the lake that reaches for four miles. It was one of the most magnificent sights I have ever seen, with the tall, slender evergreens (looked like Norway pine or maybe large spruce) reaching high above the lake on the sides of the mountains, and reflecting on the moonlit lake.

As the train neared Seattle we were all promised rooms at a hotel, where we could have baths and clean up in general. After traveling in such dirty cars—without even a chance to change clothes—baths were surely anticipated. They seemed like one of the main features of the trip.

Our train finally arrived in Seattle May 16, as did the train from Michigan. Washington Governor Clarence Martin, Seattle Mayor Charles Smith, and other dignitaries greeted us. We later discovered that the ceremony and festivities were perhaps less boisterous than what the Minnesota contingent received in San Francisco since Washington businesses were still somewhat miffed at the business California had received in supplying the new colony.

Surely you aren't Going to that God-Forsaken Country?

We didn't notice all the hullaballoo when we arrived in Seattle. We were just happy to be off the train. They housed us at the Frye Hotel at government expense. Spectators lined the streets as we walked from the train station to the hotel. We heard one man remark as we went by, "I feel sorry for Wisconsin and Michigan—they can't have many children left back there." I overheard two policemen talking as they guided us along and one said "Well, I hope they know what they are doing." Of course, we really didn't—but knowing would have taken some of the edge off the adventure!

We were so grateful when we finally reached the hotel. It was good to get into tubs and have real baths and fresh clothes. We were so dirty and so tired. We were guests of the Seattle Chamber of Commerce and were fed at the government relief station, where the "homeless and destitute" were fed. There was quite a bit of laughing over that. Everywhere we went, we were the object of stares—we certainly found out how it felt to be in the zoo. The Chamber of Commerce in Seattle arranged to have all our personal laundry done—which was more than a blessing.

Seattle's residents and businesses, while they may have been disturbed at the business that California had taken away from them, still treated us well during our visit. All the colonists were issued badges allowing free access to the city's movie theaters and trollies, and various organizations provided services such as city tours and babysitting.

We were fortunate enough to have friends in Seattle, with whom we spent much of our time. It was a rest to be apart from the crowd—but we were glad, too, when the notice was sent out that we were to board the boat—because that meant we were on our way to Alaska!

Two

Alaska Bound

U.S.A.T. St. Mihiel. chartered to carry the colonists and supplies.

We boarded the St Mihiel at the Bell Street Terminal for the ocean voyage to Alaska on May 18. We were all "eager to be at it," but that was before the ship sailed straight into the gulf storm. Although the crew tried to keep the ship in order, the washrooms and toilets (though very clean at the outset) soon became positively revolting—and all together, along with being seasick, the whole thing was just one terrible nightmare.

There was vomit on the floors, stairways, deck sides and everywhere—only occasionally cleaned up in the mornings. A few mothers who weren't too sick managed to get diapers rinsed out, but others did not. The odor from some of the state-rooms was terrible.

When the girls and I descended to the hatch to sleep, we found our hammocks cluttered with the neighbor kids' orange peels, apple cores, and vomit boxes. With our stomachs already upset from the roll of the ship and the vomit on the stairways, we were dizzy and couldn't undress. We merely pushed aside the litter, put the vomit boxes in the garbage can, and rolled into our hammocks. Once in the hammocks we slept alright, but we always rose early to escape to the decks.

Because of the horrible conditions below deck, most people tried to spend as much time as possible out in the fresh air. I loved the ocean and gazed at the waves for hours, but a stiff cold wind blew throughout the voyage, lifting the stinging sea spray across the decks.

We tried to bring our blankets on deck, but the crew adamantly refused to allow them out of the rooms, saying that too many blankets had disappeared on the voyage to Alaska with the Minnesota contingent. Even those of us who were dressed in winter clothing were miserably cold, and some had come dressed for summer and couldn't get at their baggage that had been checked through. If we went into the social rooms to get warm—we soon became sick from the overcrowded stuffiness.

We fortunately had our extra baggage, containing warm clothes. We did not consider it so lucky when our trunks had been forgotten in St. Paul, though. The trunks caught up with us at Aberdeen, South Dakota, but it meant lugging all our clothing with us after that. Everybody else's baggage went straight through to Seattle and was loaded in the ship's holds. The result was that we could get to our winter clothes, and almost no one else could.

Eating in the mess-hall below waterline was as much a trial as braving the decks. The food was greasy and unappetizing, and no matter what was served—it smelled of cabbage. Food was served in aluminum tins having several compartments. (One little girl cried and asked her mother why we had to eat in cup-cake tins.)

All dishes were washed right in the dining room and the noise and confusion were terrific. The tables and chairs slid and many times tipped completely over. I attempted to eat in the dining room just once.

After four miserable days, the St. Mihiel finally sailed into Resurrection Bay and docked at Seward, the southern terminus of the Alaska Railroad. Although we still had about 125 miles to travel, we were in Alaska!

A Creek, a Hill, & a Forty

The city of Seward on Resurrection Bay—the southern terminus of the Alaska Railroad, and debarkation point for colonists.

All of us certainly were glad to get off the ship the morning of May 22 after the sickening trip across the Gulf. We arrived at Seward at about eight o'clock in the morning and disembarked as quickly as possible.

Anxious to get to our new homes in the Matanuska Valley, we eagerly crowded onto the train. We sat there until noon waiting for freight to be loaded. The process took so long that we were sent to the town theater until the train was ready to go.

When we got back to the train after lunch we were informed that not enough tents were erected yet in Palmer to hold us. While the men had to take the train that day so they could be in Palmer the 23rd for the drawing of farm tracts, we women and children would have to wait until the 24th to go to Palmer.

Boy! Was there ever a protest! But we had no choice. All our baggage that was already loaded on the train was supposed to go back to the ship, too! But the people in charge finally relented and allowed the baggage to go on to Palmer with our men.

Staying in Seward was not so bad for me—but it was terrible for the mothers with so many small children. So many families were quite large in number and young in age—and so many of the mothers expecting more. They had enough trouble when the men were there to help—but many had as many as five children under seven or eight years of age and it was something to realize they had to go back and manage the family alone.

Many families parted tearfully that night before the train departed at 9:30 p.m. If the men had been leaving for a 3,000 mile trip never to return they would not have received more sincere farewells than those they received for a 125 mile trip and two day absence.

After our men left for Palmer, the women and children had to spend the night on the St Mihiel. We didn't want to spend any more time on the ship, but grudgingly boarded it again. There was some apprehension on the part of many women since some of the ship's crew had been drinking and many of us hated to go back alone to the boat, unprotected. But the captain stationed guards all about the boat all night so there really was no danger. Although the dangers of the voyage were past, and no one was sick, conditions on the ship were still intolerable. Everything was so revoltingly unsanitary that we were sickened at the thought of staying on the boat—and it was really too cold to enjoy staying outside.

We were only too glad to get onto the train Friday morning bound for Anchorage. That trip might have been pleasant with all the gorgeous scenery along the way, if we had not all been so tired, and dirty, and "icky" feeling. The babies—ranging in age

Anchorage in the 1930s. Ship Creek is located just above the city proper.

from 11 days on—were all upset—off feed—cross—sick—tired; and it was such a crying, quarreling confusion.

The mothers (poor souls) were too tired and distracted to handle the situation comfortably. Most of the children would not let anyone else do for them, except their mother, so those of us who did not have babies, and tried to help, couldn't do much.

It seemed as though we never wanted to hear a baby again and still we were so very sorry for them. I certainly was thankful my own three girls were old enough to understand the situation and make the most of it.

The train stopped in Anchorage on the way to Palmer and we were treated to lunch by the community. Anchorage had already welcomed the Minnesota colonists on May 10, but they graciously welcomed us.

After lunch I found a friend, Pearl Brown, who was living up in Anchorage and whose family had been our neighbors in Wisconsin. It was so good to see a familiar face—and so restful to go to her home and relax for the two hours before the time to take our train to Palmer. Even so, we were glad to be on our way—to see our new home—to be reunited with our husbands, who had drawn lots for the 'forties'—and to know who would be our neighbors.

We finally arrived at Palmer that afternoon. The community had few permanent buildings. Aside from the railroad station, a post office across the tracks, a small store and a farm house, the only other buildings were temporary ones built to house colony offices. There were the rows and rows of tents erected to house the colonists until houses could be built.

Initial planning for the colony estimated that 200 families would be sent to Alaska. However, 201 families were chosen from the lake states (68 from Wisconsin, 66 from Minnesota, and 67 from Michigan). Two more families with ties to the Matanuska Valley were added to the official rolls. One man had homesteaded in Alaska and then returned to Minnesota with his homestead application still pending. He relinquished the appli-

Northern end of Palmer. The store/post office mentioned by Margaret is at the right edge of photo, with trucks parked in front of it,

cation for inclusion in the project. The other family added to the project roll was from Wisconsin. They inherited a homestead in the valley but lacked funds to move to Alaska and improve the homestead, so were accepted into the project. Two additional families who had come to Alaska on their own to homestead, only to find the Matanuska Valley closed to entry, were also added to the project. [1]

Project officials, transients, and the Minnesota men had been working night and day at clearing and construction, but accommodations had still not been completed to take care of all of us. So those with babies were taken in by Minnesota people. Our family slept on the train, in the seats. Supper consisted of cheese sandwiches, coffee, and condensed milk with apples for the adults, and oranges for the children. Breakfast was corn flakes, milk, bread, coffee, oranges, and apples. Both meals were served "hand out" from the train.

So many of us, including colonists, some government officials, and most of the reporters who came along, had a vision of the Matanuska Colony blazing out a new life in the wilds of Alaska. But we discovered that Palmer was not in the middle of the Alaska wilderness. It was less than 50 miles by railroad from Anchorage, one of the centers for the Alaska Railroad and a busy town of 2,000, a large city by Alaska standards. There were more than 100 miles of road in the valley, and Palmer was less than 15 miles from the small settlements of Wasilla and Matanuska.

The valley already had numerous homesteads, although many were idle. Some homesteaders had given up, others had proven up on their homesteads and then left. Of the homesteads that were occupied, only about 600 to 1,000 acres were actively cultivated.

Even before we reached Seward, more than 600 people, including temporary workers and the Minnesota families, had already arrived at the site of the new colony at Palmer, which then was just a stop on the railroad spur to the coal mine at Chickaloon.

The Minnesota families arrived in Palmer on May 4 and were assigned tents in Palmer. They then waited for the drawing for farm tracts, which took place May 23, the day after we docked in Seward.

Palmer in the spring of 1935, looking northeast. Post office/store in previous photo is located to the left of the road, near the top of the photo.

Neil told us about the lottery after we arrived. By the time all the colonists arrived, 208 farm tracts had been surveyed and were ready for the lottery that would decide ownership of the tracts. Most tracts were 40 acres, although a few, because of poor soils or marshy land, were 80 acres in size.

Our men traveled all night to be at Palmer in time for the drawing of the farm tracts. After they arrived the morning of May 23, the drawing got underway. A platform was set up with a map of the valley to one side. Don Irwin, general manager for the colony; Ross Sheely, assistant manager; Colonel Otto Ohlson, manager of the Alaska Railroad; and M. D. Snodgrass, director of the agricultural experiment station in the valley, were on hand for the drawing.

Colonel Ohlson, holding a shoe box full of paper slips with tract numbers written on them, stood on a platform while the men stepped forward and drew their tracts. The men had conducted a lottery before the drawing for the tracts to determine their places in line, so the affair was fairly orderly. Arthur Hack, from Ogilvie, Minnesota, was the first to draw a tract. Even though the drawing for tracts was orderly, it still took three hours. After drawing their tract numbers the men ran to the map to see where their tracts were located, while women and children crowded around.[2]

The area was a sea of confusion after the drawing as friends and families conferred. Colonists were given 30 days after the drawing to exchange tracts with each other so they could be near friends. Ferber Bailey, who with his wife Ruth had become very friendly with our family on the train, exchanged his tract so he could be near us. Some of the tracts were also still covered with snow so their suitability for agriculture was unknown. Indeed, after the snow was gone, 11 tracts had to be abandoned and new tracts assigned.

Some Minnesota men took advantage of their early arrival and reconnoitered the tracts. By the morning of the drawing they knew what areas they wanted to live in. In some instances, if they did not draw the tract they wanted, they were able to trade for it. Some other families exchanged tracts sight unseen, only to find they had traded cleared land with house and barn for unimproved land. Although our tract was uncleared, we were quite pleased with our draw.

The morning of the 25th, we all boarded buses and rode to the camp nearest our tracts. The day before, Neil had been able to borrow a truck and move all our baggage out to our camp. Ten camps in all were established, including the camp at Palmer and the transients' camp. Our camp was Camp Roselyn, later known as camp 7, and it was about 5 1/2 miles from Palmer. The forty upon which it was located adjoined our forty. Most of the camps were quite large, but ours had only eight tents.

We in camp 7 were fortunate since the camp was on an improved tract with a house and well. We only had to go 100 yards for water. Some camps had no water facilities, and until wells

Camp Roselyn, also called Camp 7, about 5 1/2 miles northwest of Palmer. This was the Millers' home until their house was built.

were drilled, water was hauled from Anchorage in tank cars, and distributed to the camps by truck.

On that first morning we rode out from Palmer, the roads were so muddy from heavy rains that we had to give up trying to reach camp by bus. We walked the last half mile to our new "home" through calf-deep mud. There was no food in camp, but fortunately some workers there were kind enough to share their lunch with us.

That night a project official came out and discovered we had no food. He returned to Palmer for a few supplies, but didn't get back to our camp until after midnight. Meanwhile, the Baileys had a small Sterno stove, and we had our cooking pots, so we heated some condensed milk the workers had left for us, watered it down to make enough for all the children, and stirred in three jumbo Hershey bars that someone had given our girls when we left Blair. That at least gave the children something warm in their stomachs before going to bed.

Our tent was a large 16' x 24' canvas structure with plank floor. I heard that the Minnesota families found their tents well provisioned when they arrived—with beds, mattresses, blankets, food, stoves, and fuel ready for them—but our tents were thrown up in a rush, and they were dreadfully ill prepared. Our tent was furnished with bunks with springs but no mattresses, and there was no stove. Workers dropped off stoves in the middle of the night.

Although the accommodations were spartan, we certainly were glad to have our own place after sleeping on the ship and trains. We were home at last!

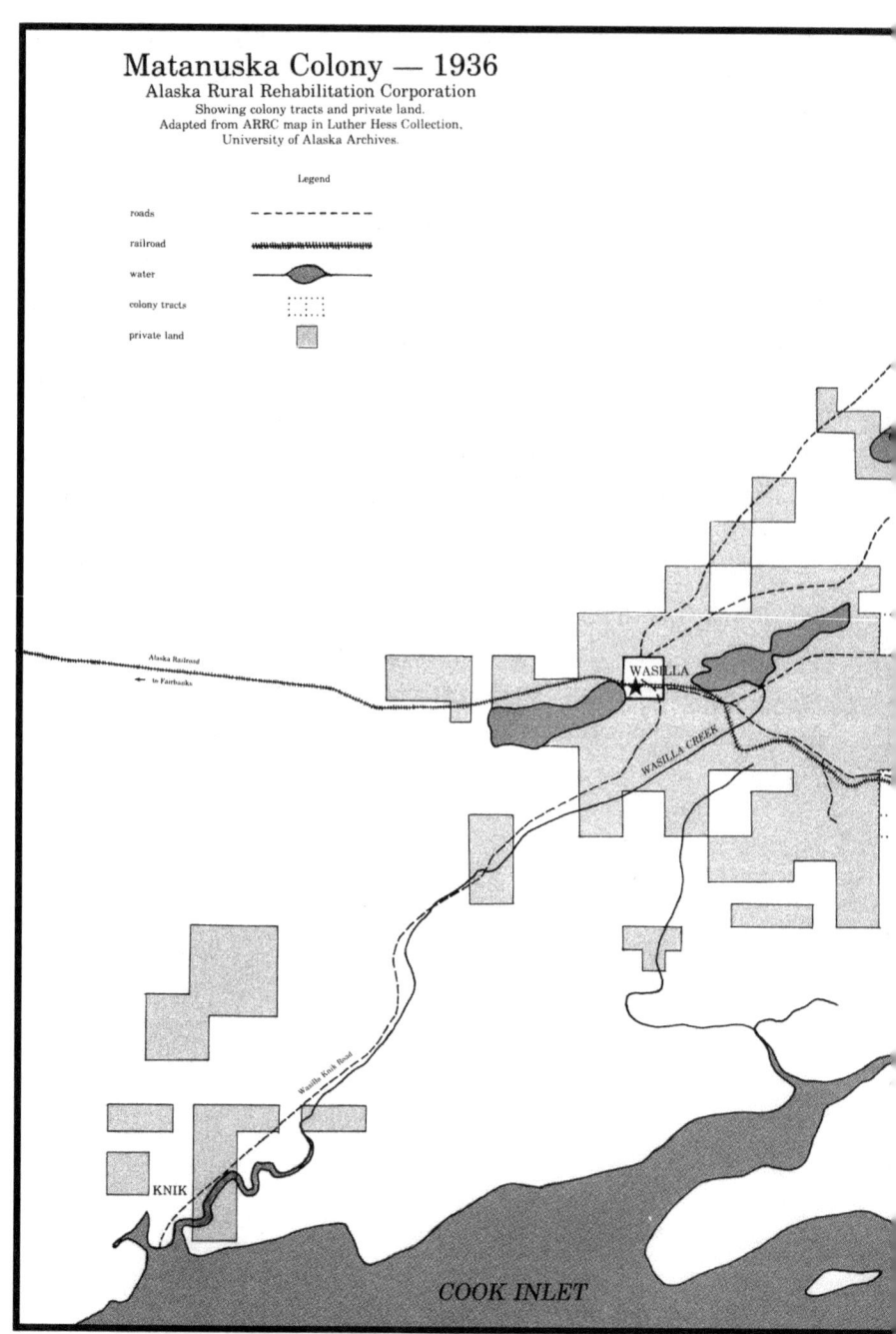

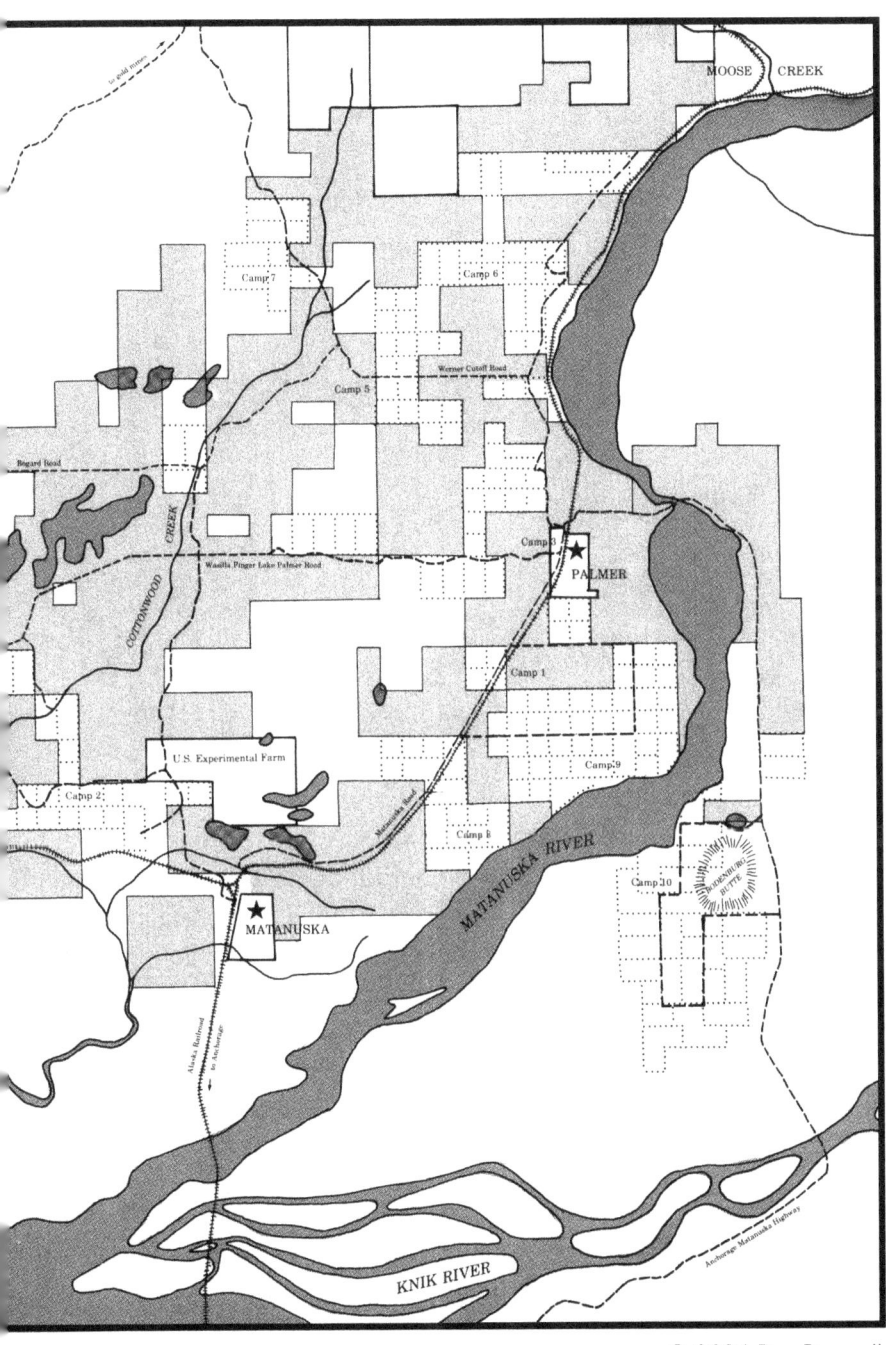

Three

First Impressions

Wash day at the Miller tent in Camp 7. I believe the small shed to the right of the tent is where the camp's sole telephone was located.

Neil brought his own tools along to Alaska and soon had our tent furnished with hand-made shelves, storage boxes, table, chairs, cupboards, even an ironing board. He had been able to borrow a truck when he first arrived in Palmer, and had hauled our baggage and household supplies out to camp before we arrived.

The tents were not well sealed, so insects and dust were a problem. Although it rained frequently, the glacial dust in the valley quickly dried, and tractors clearing land and trucks hauling supplies stirred the dust up. Sometimes the trucks stirred up so much dust that when they met on the road they had to stop and let the dust settle before passing each other.

Once the dust was airborne, the almost constant winds blew it everywhere and it covered everything. We piled dirt around the bottom of the tent to seal out some of the drafts, but

First Impressions

we had to suffer through most of it, since oil cloth to seal the tent's cracks, and mosquito netting were not yet available at the commissary.

The commissary was so poorly stocked when we arrived. It did not have washing equipment or many other household necessities, and there was no clothing of any kind. We could not even get a broom. Neil finally made one of twigs tied to a small sapling handle. It was not very efficient, but with it we got most of the big chunks of dirt out of the tent.

I was definitely unhappy with the stove furnished by the ARRC. The stoves were the cheapest, poorly made things I had ever seen. The metal in some was warped, and others like ours were missing parts.

The stoves rattled and shook whenever the wind blew, and it blew at our camp most of the time. Sometimes, parts would shake off and drop through the cracks in the plank floor. We could not even bake in our stove until the end of June, when a neighbor found a metal part in the dirt that fit ours. The Minnesota colonists were fortunate since their home state relief agency supplied them with good stoves.

I am sure we could have bought as good a stove for $10 anywhere in the states, and we heard at first that we would have to pay $40 for them. Such a ruckus was raised about the stoves, though, that the Corporation sent two of them back to Sears protesting their price and quality, and was able to negotiate an adjustment. Project officials told us better heating stoves would be available when ours houses were built, and we could keep our clunkers for heating workshops or barns.

At first we did not know the price on any of the things we received. Even if we had known, there was no accurate accounting done. Groceries were brought out twice a week and piled in one of our tents. Then the supplies were doled out among ourselves.

A record of sorts was kept of who received what, but there was no way of measuring or weighing anything. Quite a number of people complained about it, but I am sure that those in charge tried to do the best they could under the circumstances.

We finally got the tent "homey" and the girls and I settled into domestic life. Living in a tent was still an adventure and the girls enjoyed helping around the camp. Carrying water from the well was one of Janell's chief delights, even if it meant using the tricky hand-operated pump and carrying water 100 yards uphill to the tent.

I thought the trip up was going to be so marvelous, and was looking forward to two weeks of not having to plan or cook a meal. But you have no idea how good it was to have our own things and cook our own food the way we wanted it! And we surely enjoyed the life we led. Neil was so relaxed and happy.

Things at first were so confused, just settling people in the tents and trying to get the freight straightened out where it belonged. The Baileys (our next door neighbors) received their dishes soon after we arrived, but for several weeks we were the only families with any dishes. Also, there was only one family in camp with laundry equipment. So we shared our dishes with the other families, and we each took turns doing our laundry. It was inconvenient, but it certainly promoted a feeling of community.

Our family still did not have the piano, sewing machine, or heavy tool box. Rumor had it that we would not get the pianos until fall, when we moved into our cabins. But I did not see how we could get along without at least unpacking the box of our piano. All our pillows and wiping towels, one mattress, all the sheets, stockings, and old clothes were packed in there.

Life was certainly chaotic during those early days of the project. Boxes of colonists' goods were missing, and the supplies delivered to each camp from the commissary were often incomplete and mixed up. Workmen began clearing roads and cutting timber for houses, but sawmill equipment had not arrived yet. We had little real work to do.

To keep law and order in our new community, a U.S. Commissioner (equivalent to a justice of the peace), and Deputy

First Impressions

The ARRC's first warehouse in Palmer, located next to the railroad tracks.

Marshall were appointed. Residents of Matanuska and Wasilla complained about transients starting trouble when they came to town, so a "Colonial" police force consisting of 11 transients with military experience was established in July, and a temporary jail was set up in an old log cabin.

One of the force's main jobs was keeping order among the transients, but they also guarded against thefts from the many project supplies stored out doors. The railroad had a limited number of freight cars, so project supplies were unloaded as quickly as possible, often just dumped along the tracks. Coupled with the warehouse shortage in Palmer, the piles of supplies along the tracks quickly grew into mountains.[1]

The ARRC was responsible for managing the colony project, but in reality, it had little authority over the colony during the first six months we were at Palmer. The ARRC was to receive funding from FERA, but it was supposed to make management decisions about the colony independently. It definitely did not work out that way.

About the time we arrived at Palmer, the FERA was replaced by the Works Progress Administration (WPA). Harry

Hopkins, FERA's administrator, became administrator for the WPA so there was a continuity of administration.

Colonel Westbrook also shifted from FERA to the WPA and retained control of the Matanuska project. (The WPA administered the project until 1938, while funds for the project came from remaining FERA appropriations. Project officials in Palmer did not make a clear distinction between the two federal agencies, and the colony was referred to as either a FERA or WPA project.[2] To avoid confusion, FERA will be used throughout this book.)

FERA administrators made most of the decisions during the early months of the project, either from Washington, or through "troubleshooters" sent to Alaska. The at-times conflicting orders coming from ARRC headquarters in Juneau, and from FERA in Washington, D.C. were some of the reasons for confusion and early discontentment in the colony.

FERA provided the Emergency Relief Administrations in our home states with funds to supply colonists' personal needs before sending us to Alaska. Some families had as high as $400 spent for clothes alone. Besides that they were given bedding, dishes, cookware, cosmetics, oil cloth, soap, etc. However, since we were not on relief, the Wisconsin ERA spent very little on our family.

Because of the lateness in organizing the ARRC, the responsibility for purchasing and shipping supplies to Alaska fell to the California Emergency Relief Administration (CERA).[3]

CERA's role as purchasing agent for the colony created some major problems. The lack of prices for commissary goods topped the list. During those first few days in Alska, we did not know what the ARRC was charging us for supplies. Nothing was said about prices when we received our deliveries.

The lack of prices may have been due to the general confusion during those first few days, but it was also due to lack of information to base prices on. When CERA shipped goods to Palmer it failed to include invoices or packing lists, so colony officials had no way of setting accurate prices. Invoices and price lists often did not arrive until several months after the goods. Repeated requests that CERA include invoices failed to achieve results. The problem was only solved when purchasing power was finally transferred in July to the colony offices in Palmer.

Colony officials ended up charging us Anchorage prices plus freight, but many colonists still grumbled that prices were unreasonably high. When the Corporation moved to adjust prices, businesses in Matanuska, Wasilla, and Anchorage complained that their prices were being undercut and the Corporation was taking their business away.[4]

Seattle was the main shipment point for goods to Alaska, and the docks in Seattle bottlenecked most colony goods. In early June, Don Irwin told us there were already 8,000 tons of freight on the docks of Seattle waiting to get in here, with more coming to the docks as each shipment got loaded.

Part of the reason for the bottleneck at Seattle was a strike at lumber mills in the Pacific Northwest. Mill workers picketed lumber shipments at the docks, delaying ship departures.[5]

Aside from the slowness of getting goods off the docks, the major problem was goods being shipped to Palmer without any organization. The ARRC hired a man in Seattle to prioritize the freight at the Seattle dock and forward items as they were needed, but he failed to do this and shipped freight on a first in–first out basis.

Consequently, boilers for our school building arrived even though construction of the school was not planned for that summer, wagons and other machinery arrived missing parts, windows and sinks arrived for houses that had not been started, and men were unable to start work because tools were lying in boxes on the Seattle dock.

Even after repeated instructions on what to ship, goods continued to arrive without order. The dock man was finally fired and replaced with another man who dutifully shipped goods as directed.[6]

Until freight problems were solved, we colonists suffered from lack of necessary items such as tools, and sawmill equipment to mill logs for our houses. Tools were finally ordered from other Alaska cities, but sawmills did not begin operating until late June.[7]

Some of the blame for the early problems could also be placed on officials in Washington, D.C. and Palmer who were inexperienced with planning and developing farm land. During the first two months, the ARRC had no caterpillar tractors with blades to clear house sites or make roads. The engineer for the project wanted to order bulldozer blades but the purchasing agent refused. The engineer finally talked to Don Irwin and the blades were ordered, but only after considerable delay. Irwin wrote in his book, The Colorful Matanuska Valley, that, "It was not only a battle against time, but more of a battle between those who knew from training and experience what should be done and how to do it, and those who did not know and still opposed."[8]

Don and other colony officials explained the problems to us colonists, and we came to realize more and more just how tremendous an undertaking the project was—and what a wonderful opportunity for those of us who were fortunate enough to be a part of it. Of course we got impatient at times and felt things could be better organized, but the more we learned about the situation, the more satisfied we were that everything was being done as fast as humanly possible under the circumstances.

One of the first orders of business in the project was organizing a "Colony Council." FERA and the ARRC were still in firm control, but they found it convenient to form a council to act as an intermediary between project officials and the colonists—to channel colonists' complaints and suggestions, and act as a sounding board. The council was also responsible for the minor day-to-day activities within the camps.

People in each of the nine colonist camps elected one man and one woman to represent them on the council. Some of the representatives from the larger camps objected to the council makeup, complaining that the smaller camps had too much power. They demanded representation in the council be based on population, but proposals to change the council make-up were rejected.

I found myself elected to represent the women from my camp. My being elected may have had something to do with the fact that we hardly knew each other, and my last name was easy to spell. At the first meeting on June 4, I was also elected recording secretary for the council. There was a shortage of writing tablets to take minutes, and I was always scrounging for paper, however, I was glad to be secretary because it gave me a chance to hear the discussions and have the written record to refer to. Consequently, I did not have to believe all the "hooey" that was passed from camp to camp.

The council met each Tuesday night. One of the first official acts of the council was passing a resolution requiring dogs in camps not be allowed to run loose.[9] Of course, there were some at the meeting who advocated solving the dog problem with guns. Such meetings we had! A great number of people thought the representatives' chief purpose was to make complaints and demand their own way. It might have worked if any two had wanted the same thing at the same time, but there were such differences of opinion, wrangling, and gnashing of teeth.

The meetings were called to order early in the evening. The representative who talked the loudest usually held the floor until he ran out of breath, when another would begin. Sometimes, representatives with sharply differing views almost came to blows. But Don Irwin was usually able to keep the lid on, and before the meetings adjourned (often after 2 a.m.), he and the other officials were able to get their business accomplished, and sandwich in the information they wanted taken back to the camps.

While additional equipment and supplies were being shipped, ARRC workmen began clearing land, and we planted gardens and made our temporary tent homes more livable. We were told the tents would be home until our houses were constructed, which would take an estimated two to four months. Work had not yet begun on logging for house logs, so our men at Camp 7 occupied themselves cutting firewood and helping us plant our community garden. Each family in camp was allotted a half acre for a garden on the cleared land at Harold Fredericks' tract. The Fredericks, who would be our next door neighbors, drew an improved tract with house and barn, and Camp 7 was located in one of the tract's fields.

The men also fixed up a fenced pasture for the cows our camp received. The cows were not assigned to individual colonists yet, so they were a community project. We were so glad to have cows. We thought it would give us plenty of milk for the children—and from the five cows we got three quarts of milk for 13 children.

Most of us colonists visited our tracts as often as time permitted, dreaming about our future homes, and admiring the country we had moved to. We went out that first Sunday and looked over our entire forty. Our "forty" did not have the creek we had often dreamed of, but it did have plenty of hill. Near the crest of one hill we found the spot where we wanted our cabin.

I did not know if they would let us put the cabin where we wanted it or not. The Corporation planned that the four families

Pioneer Peak, on the eastern border of the Matanuska Valley

First Impressions

of each section would build their homes near enough together to use the same horses, tools, machinery, etc. At least, that was the rumor. We did not hear anything definite for a while. The only definite order we had then was that as soon as we got settled in our tents, the men were to get out the timber for our cabins. I did not know where it was to come from. None of the timber near our camp was large enough.

The Matanuska Valley is about 45 miles wide, and 60 miles long. It is bounded on the north and east by the rugged Chugach Mountains and the west by the more gentle Talkeetna Mountains. The view from our camp was magnificent! I marveled every day at the mountains. We were between two beautiful ranges which could be seen from almost anywhere in the valley. The mountains sometimes seemed to move near, then far away. At times cloud formations made them appear to be riding way above the clouds. Then again when the sun was full and clear on them, they seemed almost near enough to see wildlife on them.

In the place where we planned to locate our cabin, the mountains were even more beautiful than they were at camp. The site was higher than most of the other land around, so I was able to look out over the tops of spruce and birch and see mountains on three sides.

Talkeetna Mountains to the west. View from top of Bodenburg Butt.e

A Creek, a Hill, & a Forty

The valley was formed by glacial sculpting and river erosion. Hills and ridges along the western edge of the valley were formed from gravel dropped by the retreat of the Matanuska and Knik Glaciers. What we called "buttes," bedrock outriders from the Chugach Mountains, dot the eastern end. Elevations within the valley range from 100 to 500 feet, with land generally rising in a series of benches from the Matanuska River.

Most of the farmers praised the soil when we arrived in the valley, but its composition and thickness were really quite variable. Soils are composed of silt, sand, gravel, and sandy and silty loams. Most areas are underlain by gravel. Near the Matanuska and Knik Rivers in the eastern part of the valley the productive soil layer could be quite thick, but in the western part of the valley, about our tract, the productive soil layer was very thin.[10]

When the colony was established most of the valley was covered by virgin timber. White spruce, birch, and aspen were found on the better drained areas, cottonwoods and willows grew along creeks, and black spruce, cranberry bushes, mosses and sedges were located in poorly drained areas. The timber varied in different sections of the valley.

Where we were located, it looked to have been all burned off quite recently and the trees (spruce and birch) were not so awfully large. Our particular forty was quite rolling with lowlands that could be easily cleared, and wooded ridges. The men were all enthusiastic over the richness of the soil.

The big game—bears and moose—had gone into the mountains for the summer, but when we first came we saw moose tracks and one of the settlers who lives near the camp said he saw two bear cubs playing on the road one day last week.

Trout were plentiful. There was a chain of lakes within walking distance from our camp where the men went fishing and brought back lovely rainbows. About half a mile from our forty was a stream up which salmon ran in July. Salmon fishing was re-

First Impressions

ally the big business all over Alaska. Women in Anchorage made as high as $12 a day in the canneries.

While we marveled at the spectacular scenery, we were also impressed and somewhat dismayed by the long days. The daylight all night played havoc with us. We did not know enough to go to bed, usually not getting to bed until after midnight. Then, the mosquitoes were so vicious that we did not get much sleep. Eventually, we were able to get a powder called "Buhak" from the commissary. Buhak, when burned, drugged the mosquitoes into inactivity long enough for us to get a night's sleep.

The mosquitoes were huge. However, those first ones were rather lazy. They sounded more like a fly than a mosquito and were more easily killed. Then too, their bites did not seem to poison us as badly as the Wisconsin variety. If you did not scratch the bite, it would fade quickly, but some of the small children who could not resist scratching looked like they had bad cases of small pox. Whenever the men worked outside, they rubbed their hands with vile-smelling coal-tar mosquito dope, and covered their heads and shoulders with netting.

It stayed light all night—only about two hours after midnight being dusky. The tractors ran all night plowing. One morning I was awake at two and saw the first rays of the sun on the tent. On many nights the men played cards until midnight without artificial light.

Naturally, the weather was not always good, and when it rained, we could not see the mountains or really enjoy the long daylight hours. And it rained a great deal that first summer. The valley's average precipitation for June and July is 3.54 inches, but it received 5.27 inches of precipitation in June and July, 1935. Roads became so bad that some camps had to be supplied by tractor-drawn sledges. "Rain, rain, and more rain made the roads all but impassable," Don Irwin wrote. "In a street along the railroad, approximately 1/4-mile long, three train loads of gravel were brought in....there were places where the weight and vibration of traffic pushed the underlying mud up through the gravel."[11] The road to Camp 7 was so muddy sometimes that even the bulldozers became mired and had to be winched out.

Road crews from the Alaska Road Commission [ARC] worked throughout the summer, often 24 hours a day, to construct new roads and gravel-surface the existing roads in the valley.[12] The ARC received funds from FERA to construct and improve roads in the valley, and complete the Anchorage to Matanuska Valley highway, which was already under construction. It did not receive the funds in time to start work before colonists arrived, so it had to work around colonists and project personnel. (Approximately 79 miles of road were constructed or improved in the valley in 1935 and 1936, and the road between Anchorage and Palmer was completed in 1936, including construction of the Knik River bridge.[13])

The valley's roads received such heavy use that summer that as soon as a road was built or improved it often needed repairs. Road construction was also delayed when a ship carrying road building equipment sank on the way to Alaska.

Because of the frequent rain, our tents were often damp and uncomfortable. Outside work such as clearing land, plowing fields, or hauling supplies was even more unpleasant in the rain. Men received raincoats and pants from the commissary, but the fit was often ludicrous. I wish I had a picture of some of them when they got rigged out in their rubber clothes. It was too funny to describe. Just imagine how Mutt and Jeff would look in breeches that also fit, after a fashion, an average man. (Mutt and Jeff was a syndicated comic strip about two friends: one tall and the other short.) There too, all could not get exact fit in coats, hats, and boots. But I guess they kept quite dry in them.

Communications between our camps and Palmer was essential so a telephone system of sorts was established. Telephone lines were strung along the roads, tacked up to trees and bushes. One or two telephones were located in homes in each camp, depending on the camp's population. About 20 phones

First Impressions

The display colonist's log cabin, used as construction staff headquarters

were installed, including those in the Corporation offices—all on a single line. We were the lucky recipients of a phone at our tent. The lines frequently broke, the reception was sometimes bad, you can imagine the confusion of having 20 phones on one party line, and we were often disturbed when some one had to make a call. But the system worked, after a fashion.[14]

Even with equipment delays, bureaucratic problems, mosquitoes, and the often rainy weather, we colonists continued to work and plan. Most of us were eager to select our home sites and house plans, but many were disappointed in the designs shown us at the project architect's office. We could not design our own houses, and had to choose between five house plans, all designed back in Washington, D.C.

I was definitely unimpressed by the plans, but was sure we could make them work. The houses, called "rustic cottages" by the architect, were to be built of spruce logs sawed in a mill on three sides.

The interiors would be finished with plywood and celotex (brand name for an insulation panel). Houses would sit on spruce piling due to the lack of concrete, and they would only have partial basements. The designers estimated that each house would need 125 logs.[15]

Most of us were especially disappointed that we could not start work on our houses immediately. Our building sites had to be approved first, and then we had to wait for the houses to be constructed by the transient workers. The ARRC felt that it would be more efficient and timely for specialized construction crews to build all the houses.

The workers, most of them transients from California, were an interesting group. Any transients that I talked with had a braggy way of talking and acting, and all had about the same line. But some of them were well read, well educated men.

One red-headed man was talking to us while we waited for church one morning. I thought I was listening to a real honest-to-goodness Irish brogue, with his r's all different. It tickled the girls when he called them "goils", and talked of being "confoimed" in the Catholic "choich" -- but Neil claimed it was not Irish—just plain "Bowery."

Regular bus service was established as soon as possible within the colony. None of the transients or colonists had vehicles and the buses took men to work and women to town during the week. The buses were destined to haul children to school in the fall, but until then they were pressed into service for the Corporation.

The buses made a big loop around to all the camps, which took nearly two hours. It was a long, tiring, bumpy ride, but the scenery was magnificent, and some women and children with time on their hands took to "joyriding." That took seats from people who had legitimate business, and the drivers eventually became very adept at spotting the joyriders and kicking them off the buses.

The roads were rough and dusty, and by the time the bus arrived in town, we were so dirty and dusty as to be unrecognizable. On Sundays we took along a rag for the seats and a brush

First Impressions

for our clothes, but we still looked dingy by the time we sat down in church.

The buses began runs to town on Sundays for church goers in mid June. Church was extremely important to many of us. Most of the colonists were Lutheran, but a large portion, including our family, were from other Protestant denominations, and about 15 percent were Catholic. The spiritual needs of the colonists were definitely not neglected in Alaska. A Catholic priest, Father Merril Sulzman, was waiting on the dock in Seward when the first contingent of colonists arrived at Seward, and the Protestant clergy were also ready for the colonists' arrival.[16]

The Reverend Bert Bingle arrived in the valley May 6, two days before the Minnesota colonists. Bingle traveled to Palmer on the same train with the town's first attorney and with the man who was to become Palmer's first saloon keeper. Disembarking from the train, he announced that he was overjoyed the Lord was keeping up with the lawyers and barkeeps.

Reverend Bingle was a Presbyterian minister and served the Presbyterians, Methodists, Congregationalists and other Protestant denominations, aside from the Lutherans. He quickly set up church in a tent and had the first church service at Palmer on Mother's Day—May 12, 1935. Since his church served so many denominations, the congregation decided to name it the Palmer Community Church.

The Reverend Rudolph M. Freiling also soon arrived to start services for the Lutherans. Seventh Day Adventists and the Church of God rapidly joined the community.[17]

Colony officials assumed that at least for the foreseeable future, no churches would be built and religious groups would hold services in the community center when it was finished. Until then we shared the "community" tent on Sundays. The Catholics had their mass at 9 o'clock. Then the Protestants had Sunday School at 10 o'clock and church at 11 o'clock in the same tent. We all rode in together in the same school buses.

The Protestants waited for the Catholics—then the Catholics waited for the Protestants. I listened in on part of the Catholic service one morning and wondered why we had two churches. I thought the Sunday School was necessary for the youngsters, but the Catholic sermon was just as full of "goodness for the soul" as our Protestant one. I gained fully as much from it as I did from ours.

The project's early days were not very settling for the colonists, even for ardent project supporters like me. Sometimes we grew impatient thinking they were a bit slow in getting at work on the actual project, but I really believed when the thing was finally worked out, it would be successful if we all cooperated. My belief that the early problems would be solved did not waver, but other colonists were not so sure, and their dissatisfaction led to further problems for the fledgling colony.

Four

Sensational Reading

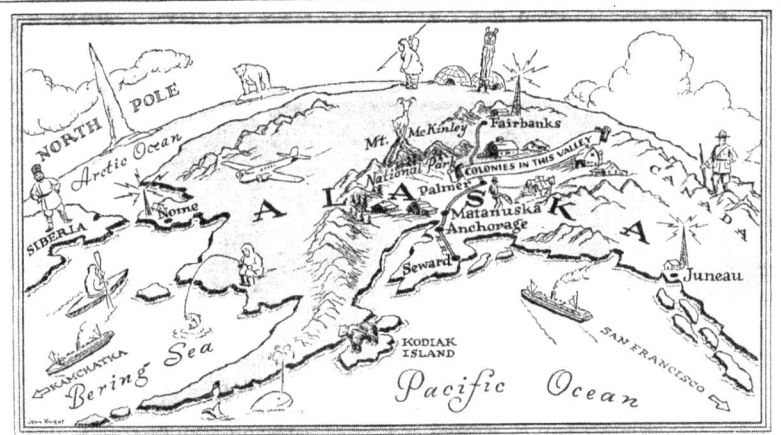

The Alaskan valley where Wisconsin, Minnesota and Michigan families will settle is shown near the center of this pictorial map.

Map accompanying a *Milwaukee Journal* article about the Matanuska Colony.

A good deal of exciting news about the colony was sent to the papers in the States. As in many other things, the newspaper men picked the "choice stuff" that made exciting reading. The fact that a majority of us colonists were content did not make good newspaper material, so that was not mentioned.

The nation was naturally attracted to the Matanuska Colony—to the romantic notion of new pioneers moving to and taming the frontier. Because of its location, the nation's press and public became obsessed about news of the new colony.

Reporters, photographers, government officials, and tourists visited the project, but most stayed only a few days, some only an hour or so. Among the seemingly endless stream of visitors the first summer were Charles Osborn (R), former governor of Michigan; Congressman Elmer Deitrich (D) of Pennsylvania, Congressman Marion Zioncheck (D) of Washington, novelist Rex Beach, aviator Wiley Post, and humorist Will Rogers.

The historian Orlando Miller wrote in his book, *The Frontier in Alaska and the Matanuska Colony*, "Perhaps the earliest and most important failure of the Matanuska Colony was that it did not for long remain a dramatic and satisfying demonstration of a recovered pioneer past. When the colony was new it was pleasant to think that once it had been possible to escape towns and panics and start fresh, but journalistic accounts of the colony before long spoiled the pleasure of vicarious pioneering....

"The press's treatment of the colony was affected not only by the disappointment of the reporters, who shared the myths and clichés of the frontier with their readers, but by an increasing general anti-New Deal slant in the press in the middle and late 1930s.... The increasingly unfriendly attitude toward the colony was not found in straight news stories from official sources but in magazine and newspaper feature articles, straining for color and written in the language of frontierism, and in editorials that showed a growing willingness to use the colony to embarrass the Roosevelt administration."[1]

Instead of self-reliant pioneers blazing homes in the wilderness, colonists were provided by the federal government with land, houses and barns, farm machinery, livestock, feed and seed, and provisions. Many of the reports from journalists covering the development "...hinted some regret that there were no covered wagons and no need to rely on ax and rifle."[2] They also picked up on a growing dissatisfaction among some colonists as work was delayed due to missing equipment, late freight shipments, mixed up orders, and other disappointments.

Colonists were called "cream-puff" and "red-plush pioneers," even by some of our most devoted admirers. Of course, many of us, who before the move had some of the most romantic visions of our new life in Alaska, now felt the term pioneer to be a little ridiculous.

We had groceries delivered twice a week, and we got furious at our orders not being filled properly, but after all—they were delivered. A nurse visited the camps as often as possible.

And they talked about "pioneering!" Everyone in our camp agreed that we all lived just as well, as far as food was concerned, as we ever did in the states.

The trouble was, the people who came in from the outside to get reports, got most of their material right in Palmer. There were so many families crowded in there. Then too, that's where all the offices, commissary, freight depot, trucks, transient camp, post office, etc. were stationed, and it couldn't help but be a turmoil. As much as most of them saw of the outlying camps was to drive in, look at the tents standing in rows, and drive out again.

Rex Beach Visits Matanuska Few Hours, Decides It's a Flop

Novelist Predicts That Most Colonists Will Quit Before Long and That Rest Will Be 'Sitting Pretty' on Lap of Federal Government; Lack of Markets Is Pointed Out

Headline from Rex Beach's article in the *Saturday Evening Post*.

A *Saturday Evening Post* article penned by Rex Beach was headlined "Beach Visits Matanuska Few Hours, Decides It's a Flop." In his article Beach stated that "The eventual outcome [of the colony] is no more certain now than it was last spring, and most well informed Alaskans predict that it will end in failure."

He went on to criticize the federal government for poor planning, for sending 200 families when the valley would only support 50 families, for the lack of markets for crops, the escalating cost for developing the project, and its high operating costs.

Beach ended the article by saying "Alaska as a whole would like to see the colonists establish themselves, but its opinion of government wisdom and its respect for government sponsored relief measures have fallen pretty low.

"It thinks Uncle Sam has made an ass of himself. Perhaps he has, but one thing is sure, the Matanuska pioneers are sitting pretty. They are safely provided for amid pleasant surroundings and they have nothing to worry about. For those who are willing to work there should be a good living.

"For those who don't care much for exercise, a benevolent government has provided a meal ticket which will probably last until the children grow up." ³

A few colonists, in some cases justifiably, began complaining about conditions in the valley. Construction on houses was behind schedule, there was division and confusion among the project administrators, and some colonists who had perhaps been promised more than they should have been by overzealous relief workers were becoming disillusioned with the project.

In a *Fairbanks Daily News-Miner* article, Don Irwin admitted that perhaps a few relief agency officials had painted an overly rosy picture of the project, and promised some colonists more than they were empowered to. He agreed that there was some unrest among the colonists, but he felt that was to be expected when such a large group of people was "...transported thousands of miles to a strange new land and consigned to temporary quarters. Despite all reports to the contrary, however, the colonists are well-housed and well-fed."⁴

Some colonists apparently didn't agree though. A "secret" meeting was held June 15, at which dissatisfied colonists (most of them living across the Matanuska River near Bodenburg Butte) aired their grievances.

Several days later, Patrick J. Hemmer and Mrs. I. M. Sandvik, claiming to represent 40 other colonists present at that meeting, sent a telegram, followed by a more detailed letter, to President Roosevelt, Administrator Harry Hopkins of the Relief Administration, Congressmen in their home states, and the governor of Alaska complaining about colony conditions.

The telegram said "Six weeks passed. Nothing Done. No houses, wells, roads. Inadequate machinery, tools. Government food undelivered. Educational facilities for season doubtful. Apparently men sent to pick political plums. Irwin and Washington officials O.K. Hands tied: Request immediate investigation."⁵

Sensational Reading

Another "secret" meeting, to which project officials were barred, but reporters managed to sneak in, was held June 21. Colony council representatives were invited to attend, and Don Irwin asked Ferb Bailey from our camp to go. Ferb wanted some support so he dragged me along.

About 100 people were there, representing about one quarter of the colonists. I felt the meeting was the most uncalled for, untruthful, disgusting, yowling mess I ever listened to. Not all the people there were unreasonable, but a core of agitators and their followers kept the easily excited colonists in a state of unrest by telling them stuff that wasn't true—misconstruing things Don Irwin said, misquoting project documents, playing up press reports that supported their view, and so on. They had some legitimate complaints about the progress of work and the like, but everything was blown out of proportion. Some of the council representatives present tried to tell the straight of the matter—but the agitators didn't listen or care.

They passed motions and made resolutions entirely contrary to those we had made at council and just generally upset everybody. The motions and resolutions couldn't be put into effect since the people had no authority, and some of the resolutions were absolutely impossible to implement anyway.

Some colonists found living on credit from the Corporation intolerable, even though they probably lived better in Palmer than they ever had back in the States. Perhaps to people coming off relief, the jangle of coins in their pockets was important. Other colonists wanted cash to buy from private businesses in the valley those items unavailable at the commissary. (This included liquor from the saloons across the tracks.)

Don Irwin agreed at the June 11 council meeting to advance families $5.00 cash each month, to be charged to their accounts, but that was insufficient for some. One of the major resolutions at the June 21 meeting demanded that all colonists be paid for any work they do, even clearing their own fields or building their own house. They also demanded that all pay be retroactive to

A Creek, a Hill, & a Forty

the time we arrived in Palmer—as if accurate records had been kept—and they threatened to strike if not paid by July 2.

They quoted sections from our settlement contracts, such as the part saying we would be provided with subsistence until our farms were in production, to mean that colonists should get paid for any work done. But they failed to include the part that any subsistence provided would be at actual cost to the colonists.

Some colonists said they had been promised jobs by relief officials back in the States, and I'm sure they never meant to make a go of farming. For the majority of us, though, the most important job was building up our farms.

There were ways to earn a little extra cash. The Corporation found it necessary to employ a few colonists at $.50 an hour to run the temporary commissaries at the larger camps, help the Red Cross nurses, and a few other skilled jobs. Colonists could also earn money by cutting firewood for settlers, and timbers for the mines at Hatcher Pass.

Complaints about the transient workers abounded at the meeting. For some reason the transients were resented, and I can't understand why. They were sent to Alaska for our benefit. But it seemed to get under the skin with some that the transient men were paid for labor that the colonists had to do without cash pay.

Some colonists didn't seem to consider the fact that they already had been given so much that they were not expected to repay. They also seemed to confuse the transients, who were paid about $30 a month plus room and board, with the very few carpenters and other skilled workers who were paid about $1.25 per hour.

I had no desire for trouble, so didn't say anything at the meeting. Anything I might have said would have gotten me in bad with the "yowlers," and they would not have listened anyway.

Dissident and dissatisfied families soon began returning to the states. Project officials stressed that no families were forced to leave, but they voluntarily left the project. During our first year in Alaska, all of us colonists were still classified as residents of our home states, and we were eligible for relief if we should return. If families quit the colony, they had to settle up their commissary debt, but the federal government paid their passage back to Seattle.

Colonists returned to the states for almost any reason: they didn't fit in, they were unhappy with living conditions, or family members became ill.

By the end of July, 26 families left the colony, and 39 families had departed by the end of 1935.[6] Families returning to the States were met by reporters as soon as they docked in Seattle and their stories, often conflicting, were printed across the country.

Senate Opens Quiz on Alaska Colony
Report Asked of Hopkins to Press Inquiry

Headline from June 22 issue of the *Milwaukee Journal*.

The *St Paul Pioneer Press* quoted one returning colonist, J. Holler as saying, "Inadequate housing, no roads, lack of educational facilities, difficulty in clearing land for crops, the too short growing season and too much rain, and grain and vegetables unacclimated to the country" were the reasons for dissatisfaction with the colony. He complained that weather was unsuited for agriculture and potatoes would not ripen in the ground, were soggy when harvested, and rotted in storage.[7]

But another returning colonist, H.O. Splittberger, said that "the government had treated them fairly and if it would stick by them for a few years, the colony is bound to be a success." He praised the richness of the soil and said if it wasn't for his wife's health, he would not have left the project."[8]

Besides the families with illnesses that had to return, it seemed to me that the people sent back fell into three categories. There were some people, apparently many from Minnesota, who were promised too much by their local relief agencies. The project was not accurately described, and after arriving, these colonists found the project not to be as they were promised. Others, mainly young childless couples, came looking for adventure and were quickly disillusioned. Finally, there seemed to be some colonists sent here by their relief agencies simply to get rid of them. This last group probably would not have fit in anywhere.

It wasn't long before press reports brought congressional response. Senators Vandenburg (R) of Michigan, LaFollette (R) of Wisconsin, and Shipstead (R) of Minnesota demanded an investigation. Reporting on Senate hearings on the colony project, The *River Falls Journal* quoted Vandenburg as saying that the colony was a "crazy experiment," in which, "the people have had to sign their lives away.... They had to agree to submit to any regulations or requirement the government authority decided to make. They have agreed to submit to any community regulation that the authorities might choose to impose as to education, health and so on. There has been established the first commune under the new deal in Alaska....These pioneers were given a one-way trip to Alaska. There is no return trip provided.... Instead of the fine fertile valley it has been pictured we are now told that it is a dry dust-swept area and the government has not provided the homes, wells, or other accommodations that they were promised."[9]

Vandenburg repeated the stories of returning transient workers about conditions in the valley. Thirty disgruntled transients returned to the States in early June aboard the North Star, saying 178 more would have returned but there was not enough room on the ship. One of the transients, William Peek, said, "... three women in the colony begged him for his identification tag, saying they wanted to cut their hair, don men's clothing and get back to the States. They wanted to get back here and work to send their families enough to break away too."[10]

Vandenburg also mentioned the first death in the colony, a 4-year-old child who died of measles and pneumonia. Many of the "howlers", as I called the dissident colonists, made a big story of it. But if it had happened in the states instead of here, it probably wouldn't even have been known, except by intimate family friends.

But there was a growing unrest and dissatisfaction towards the existing medical arrangements. We had no doctor whose work was solely to care for the colonists. There was a Red Cross nurse, Madelon De Flora, who visited camps and made sanitation and health inspections, and the transient camp had a doctor, who gave his services when notified by the nurse. Those who were seriously ill were taken to Anchorage to a hospital. It was not an ideal situation, as the transient's doctor had too many for whom he was responsible. We were assured at the council meetings that the colony would have a doctor of its own as soon as the right one could be found. But in the meantime, we had as good medical service as in most small villages in the States, and I don't believe the percentage of deaths was any greater than it would have been in a similar group anywhere else. But the papers certainly made a lot over it, as well as anything else that could possibly make sensational reading.

There were lots of slip-ups, mistakes, and poor management in certain places, but they were all just temporary inconveniences. There was no one suffering for lack of food, or milk, or anything like that. As one of the women in our camp said, "Anyone looking at any of us in this camp would know things weren't so bad." (Every one of us had noticeably put on weight.)

The colony of course had its advocates. Senator Bone (D) of Washington defended the project. In the same article in *The River Falls Journal* describing a Senate hearing on the project, Bone said the complaints about living conditions in the new project were "something in the nature of a tempest in a teapot." He praised the project as an "effort to open Alaska as a 'new frontier' by sending settlers into a valley 'rich and fertile.'" and added, "When eleven million families are living in homes definitely in the slum class in America, there is no reason to tear our nether garments because of families living in good log homes in Alaska."

Former Republican governor of Michigan, Chase Osborn, was one of the surprise supporters. He was a strident opponent of the Roosevelt administration, which he termed "delirious," and many of us assumed that after his week's visit to the colony, he would condemn the project. But he said, "I have not found one single critical or serious or sensational condition. The stories sent out about the camps by sensational correspondents are not only exaggeration, but in many instances are downright lies. Much of the discontent is purely psychological and is disappearing."[11]

Some of the journalists covering the project also reported favorably, sometimes embellishing the accomplishments of the colonists. Arthur Stringer, from the Saturday Evening Post, visited Palmer during the first summer and met Margaret. She corresponded with him after his visit, becoming somewhat of an "informant."

I was a little disappointed in his "Red Plush" article in the *Post*. He exaggerated just a bit too much. His remarks about the ex-school teacher who canned 400 cans of salmon in one day and had it stored in the piano box, and the rows of lacquered tin cans filled with cranberry, blueberry, and strawberry jams, is supposedly of me. But the truth is, he just colored up what he read in my letters about having been over to the experiment farm when six of us canned 400 cans in one day. I also never saw a strawberry all summer. But I still think he was a good fictioneer, with an admirable literary ability!

Another reporter, Arville Schaleben, accompanied our contingent to Alaska and stayed at Palmer several months during the summer. He wrote in one article about the richness of the valley's soil. His story began, "Bluebells are blooming in the woods and oats are sprouting in the fields. Things are growing in the Matanuska Valley. Summer is at hand with its abundance. "Here you plant a hard little pea seed and in three and one-half days it is kicking an infant stem out of the top of the rich earth. Oat seeds push their shoots from dirt to sunlight in four days. You lay a line of radish seeds and in 25 days you are salting their husky progeny and enjoying them at the dinner table. "Yes, things grow and grow swiftly in the Matanuska Valley. About the soil's fertility there's not the slightest doubt. About the slanting sun ray's power to give quick life and lusciousness there is no question."[12]

Congressman Marion Zioncheck (center) from Washington state, during his visit to the Matanuska Valley

Because of the confusion of voices regarding the project, the U.S. Senate passed a resolution on June 21 calling for an investigation of the project. Harry Hopkins, FERA administrator, responded, "The great publicity which has been given to this undertaking has made news and headlines out of the very ordinary difficulties of construction and administration which would normally be expected in an enterprise of similar nature, even if undertaken in a well-settled part of the U.S." He went on to give a description of the project, including history, difficulties encountered, the living conditions of the colonists, and the facilities to be provided.[13]

The FERA sent Eugene Carr to Alaska to investigate conditions in the colony. He had done similar investigative work with other government projects, and was described by newspapers as a "trouble-shooter." Around Palmer he was referred to as the "Big Bad Wolf."

Carr denied that he was replacing Don Irwin as manager of the project, but he did assume many of the managerial duties when he arrived. Carr came to the colony to investigate all our problems, and he promised results. If all the hollering howlers had been weeded out, or had gotten to work as hard on something constructive as they were at howling, probably it would not have been necessary to get someone to come from Washington to help.

He was very keen witted and capable, but he was rather brusque in his ways, and hurt the feelings of many—both project staff and colonists. He believed the project's administration was top-heavy, and started winnowing out the incompetent and superfluous staff. His first action was to fire the commissary manager, supposedly for skimming cream for his own use from the milk brought in for resale by colonists. There was widespread dissatisfaction with the way the commissary was being run, and the manager's dismissal should have increased Carr's approval among the colonists, but he stepped on too many toes, and many colonists refused to cooperate with him.

Carr was quickly followed by an investigative team headed by Samuel R. Fuller, a New York industrialist. Harry Hopkins, who selected Fuller to study the Matanuska project, said that, "Fuller will be accompanied to Alaska by a small technical staff to advise on construction projects and is clothed with complete authority to take any necessary action to bring the construction program in the valley to a speedy and successful conclusion."[14]

Fuller and his team arrived in Alaska July 14. David Williams, Chief Architect for FERA, and one of the members of Fuller's party, wrote, "Our job up here is to get the 200 families housed by winter at any cost. The rest can wait until next year if necessary. Houses, an emergency hospital, and some passable roads, and a temporary school are the main things to get done."[15]

When Fuller's party stepped off the boat in Seward, they were immediately faced with a crisis—a scarlet fever outbreak in Palmer. Although Fuller had several doctors on his staff who could have dealt with the emergency, he felt the colony's medical situation needed a permanent solution. Dr. Earl Albrecht, the recently arrived assistant to Dr. Romig in Anchorage, was pressed to accept the post of project physician, and he eagerly agreed. Three hours after being placed in charge of the scarlet fever crisis, he and two nurses were on the train to Palmer, with medical supplies to set up an emergency hospital.

The emergency hospital was a converted "temporary" community hall built by colonists and the Presbyterian minister, Reverend Bingle. Transients and other workers outfitted the hall as

Samuel Fuller (3rd from left), and his entourage after arriving in Palmer

a hospital overnight, and carpenters were still working when Dr. Albrecht saw his first patient. The hospital immediately had 12 patients, including one with infantile paralysis, and a quarantine was imposed on the entire area until the danger was past.[16] The addition of Dr. Albrecht, and the rapid way Fuller handled the situation greatly boosted the colonists' morale. Fuller also arranged for an Anchorage dentist, Dr. Pollack, to make periodic visits to Palmer to treat colonists.

Fuller reported to Washington that some of the problems with the project stemmed from the scattering of the farms. Originally the farm tracts were to be clustered around the community center, or adjacent to each other. The project plans showed a typical farm set-up of four 40-acre farms with a common corner. Houses and out-building clustered at one corner of the tract, near the adjacent farms' buildings. Such a set-up would have fostered the sharing of equipment. However, farms were scattered over a large area, with some over 10 miles from Palmer.[17]

He also explained that delays in construction were due in part to the majority of farms being cut from virgin timber, and the necessity of extending roads to the tracts before they could be developed. Fuller was confident that the project could get back on schedule. He stated that before winter set in the colonists would be housed, they would have sufficient food, fuel and water, adequate temporary medical and hospital facilities would be available until permanent medical facilities could be provided, and adequate temporary school facilities would be provided until permanent school facilities were constructed.

He concluded that the colony would be an eventual success, but the number of families would probably be reduced from 202 to about 125 due to the "...errors of selection and through the rigors of work and climate and through losses from other natural causes." He also added that it seemed probable that the colony would not be self-supporting for two to five years since the farms had to be developed from virgin forest, and "farms are spread out over so large a territory that mutual or government help is necessarily slow and difficult."[18]

Fuller came to Alaska with full powers to administer the project. Although Don Irwin may have still been the project manager, Fuller was in complete charge.

It seemed to some of us that Don was unjustly criticized, and he certainly took a lot of abuse. But he was so very gracious about it all. He never seemed to lose patience, and was so sincere in everything. I believe that every colonist liked him.

Pat Hemmer, a colonist who very early criticized the colony, was quoted in *The River Falls Journal* as saying that Don should be made "absolute dictator immediately. This colony needs somebody who can screw down the thumb and say 'Now, frog, you're going to jump and be quick about it.'"[19] But throughout the rest of the summer and into fall, Washington administrators managed the colony.

We for the most part were well satisfied that everything was going as well as possible. Of course there are agitators and trouble-makers in every group, and our group was no exception. No matter where one found a group of colonists together, he would find someone complaining. They said so much and talked so loud it sounds as though we were all dissatisfied.

But really the majority were working and getting some place. The men at the head of the colony were all fine men. They no doubt made mistakes. I don't think they realized just how tremendous an undertaking the project was.

Project officials originally estimated it would take about a year to get the colony going, but it appeared to me that it would take at least two years to get our farms established. The project had no precedence to follow, and errors were bound to take place. But if everyone could only have forgotten their petty personal grievances and jealousies, and followed when the administration was trying to lead, everything would have been OK.

Five

Settling In

Colonists continued to work and plan while administrators struggled to speed up construction, improve management efficiency, and combat bad press. The project was aided greatly by several individuals not officially connected with the project, such as Father Sulzman, the Catholic priest. Sulzman was the colony fire warden and conservation agent, and helped organize Palmer's first orchestra.

Mrs. Lydia Fohn-Hansen, home economics agent from the University of Alaska [Extension Service], taught classes in cooking, crafts, food processing, and home decorating. I'd seen a good many home economics teachers that I thought were fine, but I never came in contact with anyone of Fohn-Hansen's ability before. She could do anything, everything that could possibly be filed under the heading of home economics and do it efficiently, and all of her ideas were so practical. She even helped us with leather work, such as purses and gloves, to sell for tourist trade, so we could have a little extra money.

Mrs. Fohn-Hansen assisted in organizing a home-makers group, and she was instrumental in opening a tourist shop in the community center selling handicrafts. Her expertise extended to all aspects of home economics and even into landscaping and farm management. When wives were feeling despondent over the house plans forced on colonists by the Corporation, she sketched out suggested changes and encouraged experimentation with design variations. She helped can salmon and berries, put out a home-makers bulletin periodically, and was generally there during a period when everyone needed help and encouragement.[1]

Another volunteer vital to turning the new settlement into a community was the Presbyterian minister, Bert Bingle. Reverend Bingle and his wife made community work an important

Settling In

part of their ministry. Soon after they arrived in Palmer, Bingle supervised the construction of a temporary community hall and a playground out of scavenged materials.[2]

During the early months of the project the Bingles were one of the few families in the area with a radio, and people would gather around their tent in the evening for the news. Their radio became so popular they could not accommodate everyone wanting to listen, so they posted bulletins on current events outside their tent. They also established a small library in their tent, using reading materials donated by the Red Cross and others.

The Bingle's home was a center of activity, and when we were in town we always stopped to hear the latest news, get a bite to eat, or just rest our weary feet. Reverend Bingle and his wife were such comfortable folks. We felt as free to stop at their home as we would have at my mother's home back in Wisconsin.

Reverend Bingle also supervised salmon fishing for all the colonists. The Alaska Game Commission had ruled that colonists could catch salmon for personal consumption, and 3,500 cans of salmon were processed that summer.

Rev. Bert Bingle (second from left) and fishing crew.

Mrs. Fohn-Hansen was responsible for canning the salmon and Bingle was in charge of catching it. He would take a few of the men and run down to Knik to fish, bringing the cleaned fish back to the experimental farm where the women washed, cut, and canned it.³

Margaret (far right) and two other colonist wives with cans of salmon they processed at the Matanuska Experiment Station.

We worked in small groups of about four women, and we could pack about 400 cans a day. Of course all labor was donated, but figuring the cost of transportation, food, cans, and other supplies, Bingle estimated it cost about five cents per pound, or less than ten cents for a No. 2 can. That was the best grade of silver and red salmon, too.

Once, two men from one camp were gone from Friday until Monday, and came back so tired and with such sore hands, that they surely felt the salmon was well paid for. They cleaned, salted, and packed the salmon between tides. One time they had 500 pounds all in kegs, and the boat sank and went out with the tide, so they had to start all over again. They finally got ten gallons of salted fish for each family in camp.

Alaskans say that one develops a taste for salmon. I was determined to be a good Alaskan, so I wasn't about to admit that we were already tired of salmon. All of the work was donated, so you can see what good neighbors we had.

Settling In

The Anchorage Igloos of the Pioneers of Alaska (a fraternal organization) hosted a picnic for the colonists near Wasilla on June 23. The Railroad offered special excursion fares, and more than 400 people from Anchorage attended.

It was such a joy to see all those people come out to support us. One of the features of the picnic was a spirited baseball game between an Anchorage team and a team from the transient camp, which Anchorage won 10 to 7. The main attraction for us though was the colonists' drawing for 147 milk cows. Since there were about 200 families, and less than 150 cows, it was agreed upon before the picnic that anyone drawing a cow would share milk with those who had not yet received one. We received a brown Guernsey (not much of a looker) that we promptly named Arabella.

Some of the participants at the June picnic for colonists at Wasilla Lake.

Everyone in our camp was fortunate enough to draw a cow. Taken as a whole, our new neighbors were a fine group of people. Living as closely as we did in our eight tents, we got to know each other pretty well—not only the good qualities, but also some of the faults as well.

Our closest neighbors were Ruth and Ferb Bailey, from Lena, Wisconsin. They had an eight-year-old boy and a baby girl.

Ruth had been a rural school teacher, and was soft-spoken, sympathetic and sensible. Ferb hadn't had much actual schooling, but had been around and was unusually keen-minded.

Ferb Bailey gives Neil a haircut at Camp 7. Mardie is in the foreground with Ferb's daughter, Nona.

Neil and Ferb were as thick as mud. I never knew Neil to have such a close friend. Ferb was a natural leader, and was easily elected the men's representative from this camp to the Colony council. Through his untiring efforts to organize the saw mill crew and get the mill set up, he was appointed mill superintendent. Some men were jealous of his authority and close contact with the officials, and they tried to make life unpleasant for him, but he was the type who couldn't be buffaloed. He stood on his own feet, and granted no favors.

In the next tent were Gus and Anna Raschke from Wentworth, Wisocnsin. They had a little three-year-old girl. Gus couldn't say a single sentence without swearing, but I had as much respect for him as for anyone in the colony. He was so kindly and full of fun—a good worker and broad minded. His wife was a bit self-conscious and hard to get acquainted with, but the more we knew her, the more we liked her.

Settling In

Walter Maningen and his wife, Vivian, lived in the next tent. They were from Champion, Michigan. The night we landed in Palmer, Walter was so drunk when he came to the train to meet his wife and baby that he couldn't walk straight. We were all disappointed when we found him in our camp, but we quickly changed our minds.

Walter only drank around other men, and none of the men in our camp were drinkers, so his problem was controllable. He was such a good worker—no matter what he was asked to do, he did it quickly and efficiently. He never swore or talked rough, was lovely with his family, and he had a keen though somewhat droll sense of humor.

Both he and his wife were very bashful—they were just kids really. Walter was about 20 and Vivian was only 18. Perhaps Walter's youth elicited the parenting instincts in us, but I believe we or any of our neighbors would have fought a pitched battle if anyone outside our camp had criticized him, we liked him so well.

Harold and Fanny Johnson, from Houghton, Michigan, were next. They had a three-year-old boy. Their tract adjoined ours, and we couldn't have asked for better neighbors. Harold was a very quiet, clean-cut person—the type of man who loved horses so much that all his spare moments were spent with the team. He was a very fine worker, too.

And Fanny was such a peach—she was the kind that "mothered" us all. She had a low, slow way of speaking, and was so sympathetic with all our little difficulties. She was one of those clean, clean Finlanders. Her tent was always spotless—with rugs on the floor to hide the splinters, an embroidered cloth covering the sewing machine, and an old cane rocker making the tent so homey.

Clarence and Alida Greene, from Hancock, Michigan, were the other neighbors our forty would corner with. They had only been married six months before coming to Alaska. Clarence was a little bit of a fellow, not as tall or as heavy as I was, but he was the biggest eater in the entire camp. He got a lot of ragging about that but took it in good humor and gave back as good. He hated his name and insisted we call him "Bagsy." Alida was still so newly-married that everything started, stopped, and kept going to the tune of Bagsy. They both were kidded unmercifully, but they were a lot of fun.

Martin and Margaret McCormick from Michigan were married when she was 18 and he was 16. They had three boys—one boy twelve years old, and twins who were ten.

The McCormicks had their good qualities, but were both braggy and swaggery. Still, a lot of their bragging was backed by results. In spite of the fact that he knew it, and didn't hesitate to tell you, he was one of the best men on a tractor among either the transients or colonists. He was considered a "key" man and was in charge of the tractors on the colonist construction force. We got rather tired of hearing about it, but he was more or less an authority on everything. However, I should be loyal—he was always loyal to me as representative, and I appreciated it.

The last tent was occupied by Theodore and Leonea FitzPatrick, from Roscommon, Michgan. They had three children and a grandpa, whom I was quite fond of. He was not very old, but Mrs. Fitz wouldn't come if he didn't. He helped with the housework, and took care of the kids. Mrs. Fitz was fat and jolly, but very strikingly pretty. Fitz himself was all right, but he couldn't make a decision for himself. He was merely McCormick's shadow and yes man.

In the house lived Mr. and Mrs. Fredericks, from Sturgeon Lake, Minnesota. (Albert and Audrey, but they always seemed to insist on being called Mr. and Mrs.) He drew an improved tract, with house, barn, grainery, blacksmith shop and equipment. The house even had furniture and dishes when they moved in. They were always unusually nice with me, and we got along well enough, however, peaceful relations depended on holding our tongues when around Mr. Fredericks. He was another who thought he knew everything, and you couldn't tell him anything.

Settling In

He also felt he was not treated as part of the group—a situation arising from his luck of the draw. Since he already had buildings, he wasn't included in the construction discussions, which naturally were the main conversations among the men.

Then, when Ferb organized the crews for the sawmill at Camp 5, one man had to be left at camp to take care of the stock and the community garden. That fell to Mr. Fredericks, since he owned the farm, so he had even less in common with the other men. His wife was as human as she dared to be, hardly saying anything without asking Mr. Fredericks first. (She always called him Mr. Fredericks, too.) They were just a young couple, possibly 23 or 24, and had two children.

Our family rounded out the community, the luck of the draw certainly giving our camp a varied assortment of personalities. Perhaps it was because ours was the smallest camp in the colony, but we became very close-knit, and colony officials praised our camp as being one of the best behaved and maintained camps in the valley.

We received our first mail from the States, including letters from my mother and several other friends, about the first of July. For the most part the letters were tremendously uplifting, but quite a few people wrote as though they were communing with the "dead and gone," as though we would never see them again, and we certainly didn't feel that way. Mail was a great lifeline for colonists, making us feel less isolated from the outside world. That first year, we colonists often received mail from perfect strangers, who had seen our names in the papers and wrote to encourage us or send small gifts.

Mail arrived in Palmer only once a week, corresponding with the steamship schedule. Likewise, once a week letters and packages were shipped south. There was no rural mail delivery, so mail day was a crowded and exciting day in town.

July first was also the day the post office's name was changed to Palmer. Before that, the train station's sign welcomed people to Palmer, while the Post Office sign welcomed them to Whar-

A Creek, a Hill, & a Forty

ton. There were some efforts to change the town's name to Valley City or various other names, but Palmer prevailed.

By July the house construction program was organized and in progress. Five portable saw mills were set up,[4] one at Camp 5 southeast of Camp 7. There was no sufficiently large timber near Camp 7, so Neil and the other men from Camp 7 worked at Camp 5 to harvest timber for houses. Leonard Herried, whose land was adjacent to the saw mill, gave Neil permission to cut our timber from his land.

Portable sawmill that was set up at Camp 5 to harvest house logs.

Because Neil was from a different camp, his timber was scheduled to be cut last. But before it was Neil's turn, one of the men from Camp 5 ran short of logs, and officials went in and helped themselves to Neil's timber without so much as a "by your leave!" For all his work during June and July, Neil barely got enough logs for piling and sills.

Fuller quickly realized there was not enough large timber on colony lands to meet the project's building needs. So the house plans were adapted to frame construction, and a rush order was put in for lumber to complete our houses.

We were told that we who lived in camp 7, and some in the other camps, were to have frame houses. We waited all summer to be told where to go to get our timber, then they finally decided it was not to be found anywhere. It was certainly a great disappointment. The house plans were designed for logs, and Neil commented that our house would look more like a chicken coop than a house if logs were not used.

But, we couldn't do anything but adjust to the new plans and continue working. Project officials finally told us we could select our own building sites, so we started clearing. Our building site stood on a knoll, with spruce and birch to the north and east for protection from the prevailing wind. We had lovely long cleared slopes to the south for warmth and a garden. (Our "forty" didn't have the stream we dreamed of, but it had plenty of hill.)

From our knoll I could see the Chugach Mountains and the Knik Glacier to the south and east, and the Talkeetna Mountains to the north and west. The Chugachs are by far more beautiful mountains. They rise to such towering snow-capped peaks, with the lights, shadows, clouds and colors constantly shifting. I could not begin to describe how very, very grand and beautiful they were to me.

Fuller realized there was little time left to complete our houses by winter and not enough workers to complete them. In early July colonists were told to stop all work except on our own houses, and 225 additional transient workers were requested from California.[5]

Adding to official worries about construction was the fear that transient workers would quit. One reason for the transient workers' dissatisfaction was their low pay. They were unhappy that they only received $1 per day (plus room, board, and transportation costs to and from Alaska) while the colonists they worked with received $.50 an hour.[6]

Just in case a major portion of the transient workforce quit, Colonel Ohlsen, president of the ARRC board of directors and also chairman of the Alaska Railroad, agreed to supply ARR personnel to complete construction of houses if necessary.[7]

The construction season was quickly disappearing, so in late July Fuller tried to speed up the program by proposing that frame houses, prefabricated at the main saw mill, and with a single floor plan, be constructed. These houses could be easily added to in the future if necessary. The basic unit would have been 32 feet long and 15 feet wide, divided into three rooms. Project officials estimated that such a house could be erected in 25% less time than a standard frame house, and 33% faster than a log house.

Many colonists took a quick look at the new house plans and declared them to be "sheep sheds," not fit for human habitation. These same people had loudly complained about the original house plans, but faced with yet another change, they adamantly refused to even consider it. At the July 23 council meeting, we voted unanimously to reject the plans. We were told we could go ahead using the original plans, but we had to hurry! Even if construction went according to schedule, the last house would not be finished until October 28.

Every family in the colony dreamed of how their finished homes would look, and each family was allowed $285 for furniture. We all made up lists of the furniture we wanted, and the Corporation ordered it from Montgomery Wards or Sears and Roebuck. Allowing for freight, our furniture allotment was reasonable. (The *Seattle Post Intelligencer* advertised new sofa and chair sets for $49, kitchen ranges for $69, console radios for $50.) The Corporation also received sizable discounts since the furniture was ordered in bulk shipments. Even so, many families went way over their budget. Some families wanted to buy pianos and flatly stated that they would not accept used ones. They had to have new pianos!

The furniture allowance was a loan from the Corporation, and we decided to make our own furniture to keep our debt low. Neil had some birch lumber for furniture sawed at the mill and set aside to season, but until our house was finished, we planned to make do with the furniture from the tent. We did have to or-

Settling In

der a kitchen range and washing machine, and we were advised to buy a good radio to get us through the long, isolating winter. Anchorage had one radio station, but good radios could pick up stations from the States and overseas.

We were delighted when our piano was finally delivered the first week in July. The piano has been stored in the warehouse at Matanuska along with many other colonists' belongings. The girls were so happy to have it. So was I, as it gave them something to do in the long idle hours. The valley was still under a scarlet fever quarantine, and both Janell and Priscilla came down with mumps in July, so it was a terribly trying month.

I was also happy to see the piano for another reason. Most of our linens were packed inside the piano case. Of course, the Matanuska dust had infiltrated the case, and everything was filthy. Washing conditions for all the colonists were poor. Our family did not even have a wash tub until July, so I had to borrow washing equipment from neighbors. Only three of the eight families in camp even had tubs. I certainly longed for the boiler and wringer I had left in Wisconsin. Some of the women nearer to town who brought their laundry equipment were even able to earn pin money by taking in the project staff's laundry.

When the commissary finally got laundry equipment, I bought three tubs, plenty of clothesline, and a board. We were all supposed to get necessary items such as laundry equipment when we first got to Palmer, but we arrived at the same time all the supplies did, and it took time for the commissary to get organized and stocked.

Although we had to contend with delays and disappointments, and less than ideal living conditions, we and many others in the project were pleased with our new lives. Life with us was very unruffled and peaceful then. The girls thought it was the "only life." Mardie said she could not see any reason for ever having any more room.

I even had a "frigidaire." The frost was still only a few feet below the ground, so Neil dug a hole under the tent floor, lined it, and made a dumb waiter to fit. It kept the food as cool as some ice boxes.

On July 24 Colonel Leroy Hunt arrived in Palmer to assume management of the project from Fuller, who then returned to Washington. Hunt was on temporary leave from the U.S. Marine Corps, and David Williams, FERA's supervisory architect, probably expressed the opinions of most of us when he said, "If a colonel of the Marines can't run it [the project] -- then it ought to sink!"[8]

Hunt had several military assistants, and rumors circulated that the colony had been placed under military rule. However, his administration did not function under military regimen, and Hunt was viewed with respect and admiration by most of us.

The morale of the colonists improved greatly under Fuller's reorganization. With Hunt in complete authority, and everyone here assured that Don Irwin's relation to us remained the same, and some of the folks who were dissatisfied went back, affairs seemed to begin clicking and running smoothly.

Hunt was a fair minded man, and was anxious for construction to proceed at the most rapid rate possible. When he talked to any of us he "put his cards on the table" and gave us credit for having enough intelligence to understand the situation. As a result everyone was in better humor.

By August things were beginning to speed up and take form. It was a new thrill every day, at least for anyone of my temperament. (I don't know when I had a more interesting summer.)

Additional groups of laborers had arrived: 75 transient workers by July 24, another 125 transient by August 1, and 150 Alaskan-hired carpenters during July and August.[9] The lumber for the frame buildings was also received in July.

Most of us in Camp 7 had our houses under construction by the beginning of August. One man had the framing up, waiting for sheathing. Carpenters were expected to arrive soon to complete construction, and everyone in camp expected to be in the homes by September 1. The house interiors would not be fin-

Settling In

ished by the time we moved in, but fall was fast approaching, and we all wanted to get out of the tents as quickly as possible.

Four well drilling outfits arrived at Palmer in late July and immediately began drilling to provide at least one well at each camp. (Drillers were busy from then until spring 1936 drilling wells for the colonists and the community facilities.)

Hunt could not tolerate administering the project from behind a desk, and he was out about the valley on most days actively assisting with and overseeing the project's progress. He got around to the different camps several times a week, and on one of his visits out here he brought along a man with a camera. Of course they came with a camera, on the day I was dressed in Neil's panama hat, a red striped sweater tucked into a pair of jersey pajama pants, which in turn were tucked into high boots!

My clothing situation would have been ridiculously funny if it was not so serious. I had worn out anything I had for work clothes, and I never had much. Clothing and many other necessities were not yet stocked by the commissary.

The ARRC assumed we were supplied with all the basic necessities by the state relief organizations, but many families were not adequately outfitted. Each family wrote out lists of clothing needed, which the Corporation sent back to the state relief agencies, but there was no assurance the relief agencies would provide additional supplies.

I had one good pair of overalls, but no shirts. I kept my corduroy slacks and wool middy for very special occasions like trips to town, or council meetings. The rest of the time I wore whatever was at hand to cover me adequately. My boots ripped down the back and I turned cobbler and sewed them up. But no one cared much about appearances. We were all in the same boat. No attempts were made to "keep up with the Joneses." That was such a comfortable feeling after having been school teachers' family and having to do "what's expected of teachers' families."

Although the Wisconsin Relief Agency did not provide us with additional clothing, it did send us an unexpected surprise in August. All the Wisconsin families received 23 cans of beef for each family member, several boxes of rice, mattresses, material for sheets and bath towels, a quilt and several wool blankets. All of this was free and greatly appreciated.

The large mosquitoes that were here when we arrived and which were not so troublesome were soon replaced by a smaller, more vicious variety. My concept of Satan and all his minions was forever altered by the appearance of those little striped devils. It was impossible for the men to work without nettings over their head and shoulders, and heavy gloves on their hands.

Silk stockings still look nice and felt more comfortable inside, but even cotton stockings were not thick enough to protect from mosquitoes, and it was only very special occasions indeed, that called for silk stockings. The rest of the time, heavy stockings and low-heeled oxfords or boots were "what was being worn by the smart set."

In between preparing our house site, washing, gardening, and doing other chores, I, like most other colonist wives, picked and preserved berries. Highbush and lowbush cranberries, currants, raspberries, and strawberries could be found in the valley. I never found any strawberries or raspberries that summer, but I did find and harvest some currants and lots of cranberries.

By the end of July we had over 10 pints of cranberry jelly preserved, more berries ready for canning, and I was out of canning tins. Luckily, the commissary had canning tins in stock by that time.

By late August I had put up about two dozen 2 1/2-size tin cans of highbush cranberry jam. Some of it I mixed with grated orange, some with prunes, dried peaches, dried apricots, and dried apples. I also had some mixed with vinegar and spices. If we picked the highbush cranberries before they were too ripe, they jelled even more easily than currants. There were many lovely currants throughout the valley, but not where I could get them conveniently, except the black ones.

Settling In

The lowbush cranberries ripened rapidly. There were so many on our place we could hardly walk without stepping on them. They looked just like unripe blueberries, and we thought they were.

Fanny Johnson, our nearest neighbor, found some ripe blueberries, and we first thought they were the same. So instead of a carpet of blueberries we had a carpet of cranberries (just a change of color scheme.)

The lowbush cranberries keep without canning. Folks picked them, put them where it was cold, and used them as needed. If they froze it did not matter. I, and probably other women in the project, received more pleasure from harvesting and preserving berries than might be expected.

It was very rare when the commissary provided fresh food, so we lived out of tin cans to a great extent that summer. Some people felt tin cans should be the colony's symbol. How I looked forward to the time when I could have the tin cans full of things I had raised myself.

Arabella turned out to be an excellent milker, so we had milk and butter regularly. When we got a new churn in August, it was a real treat making butter. We had made it with egg beaters before, but using the churn was a special thrill.

We started right after supper. The girls took turns until bed time. I was making jelly so when the girls went to bed, so I let the cream stand until I was through. Then I turned the churn for another half hour or better before I could see any signs of butter.

But some way those chores did not seem tedious. The butter has so much finer flavor and there was so much more joy in eating it, when we knew it was made from cream from our own cow, made in our own churn, by ourselves.

By the first week in August, we had our building site cleared, and working by ourselves with a horse and scraper we began digging the cellar. The gravel we dug through was not very well sorted, and we often came across large cobbles.

Neil and Margaret with a horse and scraper, working on their cellar.

The hill on which our house was to be built was rather steep, and never missing an opportunity to be frugal, I carried the cobbles to the bottom of the hill and started a retaining wall. Then we filled in behind the wall with the excavated gravel to make a more gradual slope.

We were so eager to finish the cellar, we didn't even want to stop to fix meals. That duty was delegated to Mardie, with assistance from Priscilla. Mardie became a very accomplished cook, as long as dinner came out of a can.

Of course, Neil and I daydreamed about the finished house as we worked on the cellar. I was looking forward to the day when I could walk into my cellar and view all the jars of various colored fruits and vegetables. We were over half way down when it occurred to me that they'd all be in tin cans. It was quite a blow! How I missed my glass jars.

The cellar excavation and retaining wall were pretty much completed by August 15. We only dug a partial cellar, but we did add a trench under the bathroom for plumbing of a bathtub. I felt we must have a bathtub, even if it meant hauling water to fill

Settling In

it. Soon after we were finished excavating, crews arrived to dig holes for the house piling.

Most of the other houses in our area were taking shape slowly as materials arrived from Palmer, but our next door neighbor's house was being readied as quickly as possible. Fanny Johnson was seven months pregnant and had become seriously ill, so carpenters were racing the stork to complete the Johnson house.

Fanny became partially paralyzed in mid-August. She couldn't be treated in Palmer, so she was rushed to the Anchorage hospital to recover. The doctors were able to save Fanny, but the baby didn't survive. Most of the people in camp 7 were real troopers and we had become very close, so the tragedy was a real blow to camp morale.

Neil and I expected carpenters to start on construction of our house any day and we completed the plans for the interior. We were told we had to follow the plans for the house exterior, but as long as no additional time or material was needed, we could change the interior to suit ourselves.

There were four families in this camp using the same general plan for the exterior, and no two houses were planned alike inside. I had the house so vividly pictured in my mind as to what it would look like inside that I was sure when I really saw it, I probably would not know my way around, it would be so different. Anyway, I had an awfully good time thinking about it.

I cannot describe my emotions and "imaginings" as construction of our home progressed. I could not help but think of some of the tumbledown, forsaken homes throughout the world, for which someone, at some time or another, had such high hopes, and such happy anticipations.

The scarlet fever quarantine was finally lifted at the end of August. Children, who had been restricted to their camps, could now move about the valley. Several 4-H clubs had been formed in Palmer, and Mardie and Priscilla went to town right after the quarantine was lifted to enter aprons in a 4-H sewing competition. Sears and Roebuck had donated fabric scraps to the 4-H clubs, and the sewing competition kept many girls busy during the quarantine. Mardie and Priscilla each won first place in their divisions. Priscilla saw her apron on display and said looking at it just made her tired all over—she was so sick of working on it. But she was pleased to win a ribbon. (While 4-H was probably the most active youth organization in Palmer in 1935, other youth groups were forming. Boy Scouting was active by the end of September, and a Camp Fire Girls by early spring of 1936[10])

Neil had applied to the Territorial Board of Education to teach, and on August 28, Mr. Thuma, the superintendent for the Wasilla school, stopped by and announced that Neil was to teach seventh and eighth grade in Wasilla. It would certainly be a hardship to teach in Wasilla, but the $160 a month salary was so inviting. Thuma recommended that we move to Wasilla for the winter because of the poor road conditions between there and Palmer. We hated the thought of leaving our new house, but if driving conditions did get unbearable, we were resigned to move. The Johnsons could watch Arabella for us, and we would return on weekends and holidays to work on the house.

We found a two-room cabin in Wasilla we could rent for $10 a month. Until we needed it, two other school teachers were living in it while their cabin was finished. Teachers' quarters were in rather short supply. One new and rather bewildered teacher at Matanuska had to take up residence at the jail until more permanent quarters could be located.

With the end of summer, Palmer saw the birth of a hometown newspaper. Jack Allman published the first issue of the *Matanuska Valley Pioneer* on August 22. Allman began his paper stating that, "Mark Twain once said that no town ever takes on the status of a city until it has a newspaper, and the citizens enough trouble to fill its columns."

Jack was a sports and fiction writer, itinerant newspaperman, and Alaskan sourdough who had joined the European press corps during World War I but never made it back to Alaska. He

Settling In

returned to Alaska with the construction workers and quickly persuaded the ARRC to sponsor his newspaper. The ARRC provided him with a tent, mimeograph, and operating funds. With his limited funds and one-man staff, Allman produced a weekly paper from four to eight pages long.[11]

He stated that the purpose of the paper was to keep the colonists and other residents of the valley informed of all going on in the valley and the progress of work on the project. The *Pioneer* related social activities in the valley and had correspondents in each camp. It also reported on current events outside the colony, printing articles on national issues, activities in Europe, and the confrontation between Ethiopia and Italy.

Allman was an ardent promoter of the colony and wrote editorials defending the project. He wrote a lengthy response to Rex Beach's article criticizing the project, saying among other things, that, "Beach, of course, is well known for his fiction with an Alaska background, but we feel that this last is by far his best efforts in this field." He also defended the project with articles and letters to national publications such as *Time Magazine*. The *Pioneer* was also sent to territorial and federal officials, FERA administrators, and other people interested in the project.[12]

The *Pioneer* reported that by the end of August the project was making rapid progress and Palmer was beginning to lose its temporary boom-town appearance. The Trading Post (new name for commissary) was finished, and the warehouse and powerhouse were nearing completion.

Work was starting on a permanent hospital and the community garage, and excavation had begun for the community water system and sewage disposal system. There were 117 houses under construction, and 41 wells had been drilled. The *Pioneer* also heralded the construction of the town's first hotel, the Valley City Hotel; the opening of a branch of Koslosky's Department Store from Anchorage, selling groceries, clothing, and hardware; and advertised the services of Palmer's first attorney, Almer J. Peterson.[13]

Six
Under Roof

The first phase of this colonization project was coming to a close. As with the birth of any child, the first acute pains were over, the feeling of relaxation was upon us, and while we realized there were plenty of worries ahead before the child reached maturity, still and all it seemed to be quite alright and taking nourishment with only occasional "colicky" spells.

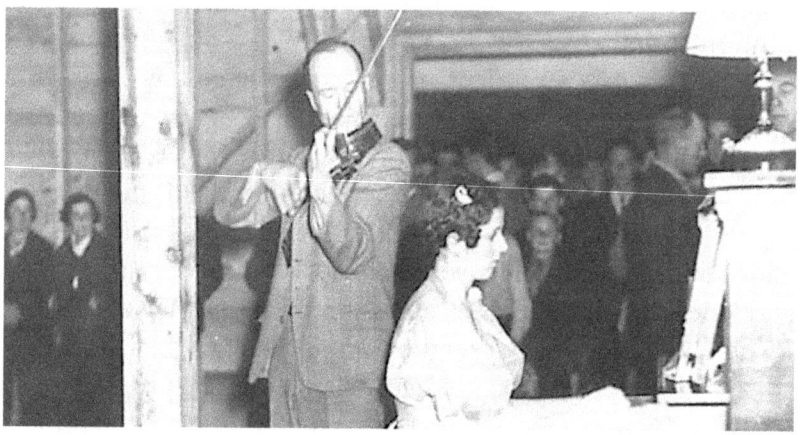

Harriett Malstrom (seated), a Seattle singer and pianist, entertained the colony in September, accompanied by local musisicans

The social event of the season was held the evening of September 3 in the newly-finished Corporation warehouse. Singer Harriett Malstrom, a Seattle "radio and dramatic star," who was touring Alaska, adjusted her schedule to perform the first concert held in Palmer. After Malstrom's performance, a dance was held, with music provided by a group cobbled together from local talent. The *Valley Pioneer* reported that over 400 people attended.[1] Attendance might have been higher had September 3 not also been the first day of school.

We didn't go down because we couldn't feature keeping the girls up until a dance was over in order to get a ride home. But

Under Roof

everyone said it was very nice—well attended and very orderly. It included colonists, settlers, transients and officials.

The start of school brought an appearance of normalcy to the valley. Since Neil was teaching, the whole family started school except "Ma." All three girls were tremendously excited. Mardie was a freshman, and as it turned out, the only colonist at the Wasilla high school. She came home saying that in order to rate at all, she had to have a wristwatch, a twin sweater set, a boyfriend, and go to all the Saturday dances, and to the roadhouses for lunch. She was quite willing to insist on the watch and sweater set, but I wasn't sure she had reached the stage in her development when a boyfriend and all the trimmings were essential. However, that sort of thing seemed highly contagious in girls of her age.

Priscilla was overjoyed to have her "idol and model," Dad, for a teacher, and Janell had a very acute attack of being angelic the first week after school started. I held my breath in fear of the reaction, but it certainly was a joy unsurpassed while it lasted. I came home from town one day that week about 4 o'clock and found she had washed the dishes, set the table, peeled the potatoes, and straightened up the house. She made her bed every day, washed her face and her neck and ears, and didn't argue all week. It was heavenly.

Neil and the girls rode to Wasilla by bus—a 13 mile ride. At first the Territory told Neil he was to drive bus, and teach seventh and eighth grades, but then the Corporation couldn't release the buses to the school. They really belonged to the school, but had been loaned to the Corporation all summer and were still being used to transport people to work. So when school started, the buses first delivered men to work, and then picked up the children and took them to school. In the afternoon they brought home the children and then picked up the men. The $40 per month driving bus would have added to Neil's salary was not exorbitant (considering that we were in Alaska where there was a reputation for high salaries) but it looked like a million to us. At least with his teacher's salary we could start paying on our debt to the Corporation.

Our debt was relatively low, but some families were not so frugal. To combat rising commissary debts, budgetary allowances were worked out for each family. Commissary allowances were determined depending on each family's size, ranging from $45 for a family of two, to $125 for a family of 13. Our allowance at the commissary was $85 per month. I averaged about $50 a month so I was not worried but what I could live under the amount.

The allowances were high then to allow ample provision for getting our winter's supply of case goods and other supplies. The commissary was finally becoming adequately supplied with winter clothes and case goods. In August, as people began thinking about winter, a food "shortage" of sorts developed, and families were limited to six cans of any one item until stocks could be increased. I don't think there was any real shortage, but once the rumor got started, the run on the commissary began.

After our winter needs were taken care of, the family allowances were to be reduced. Some families spent way out of proportion, though. Two families spent $400 each during July, one family spent $285, and 58 spent over $100. So the management had to tighten down. I couldn't imagine what people could have done with so much groceries (they weren't the largest families) unless they were storing them up. One family who went back turned in $200 worth of groceries.

Colonel Hunt warned us all to live within our budgets.[2] Besides warning of some colonists' mounting exorbitant debts at the commissary, he also told of one group of angry colonists that had returned coffee to the commissary because it was not the brand they ordered. The commissary was out of the brand they wanted and delivered a substitute. He said such practices and attitudes must stop.

Hunt emphasized that, "...Credit has gone to the heads of some of you, and the only way to keep that situation from growing out of hand is to clamp down hard.... The presents are all gone. The Christmas tree has been thrown out onto the brush pile, and from now on you must manage to live within your budgetary allowances."

Construction of the school house in Palmer had not been started by September, so temporary measures were initiated to teach our children during the 1935-36 school year. There was no place in Palmer to hold classes, so until something more definite was provided, children from camp 7, camp 4, and part of camp 2, and all adjacent settlers' children went to Wasilla. The children from camp 8 and part of camp 2 went to Matanuska. The rest of the children were given books, and teachers visited the homes twice a week for recitations.

We still hoped to get our $150,000 school house in Palmer, but Colonel Westbrook in Washington advised against it. He wanted four smaller "temporary" schools built in different parts of the valley that winter. However, we feared the temporary schools would become permanent, and that the central school in Palmer would never be constructed.

People were generally dissatisfied over it. Many of them made their decisions to come because of the assurance of the fine school. As long as there was still some assurance that the good school would be built eventually, there was no open protest, but I was afraid of what would happen if they took that away. The school was to be an outright grant to the Territory, and many colonists planned for it. The settlers did too.

As far as we personally were concerned, no promises of any sort were made. But it seemed some folks were promised so many things. Then one thing after another fell through that they had planned on, and some colonists became rather discouraged. If officials took the school away it would appear as though the government had lost faith in the project.

The initial construction phase of the project—getting the colonists housed—was coming to a close, and many Corporation workers started work on other ARRC facilities such as teacher and staff housing, as well as finishing work on the trading post, hospital, and other Corporation buildings.

Families in the Bodenberg Butte area, whose tracts were across the Matanuska River from Palmer, and in some cases were over nine-miles distant, began work on their own community hall. With ARRC approval and using volunteer labor and locally-harvested logs, they built a 28'x50' hall. According to a *Valley Pioneer* article, the only supplies provided by the Corporation were windows, roofing and flooring. Completed in mid-October, it was pressed in to service as classrooms for local school children.[3]

Gym/community center and school house under construction

We finally did get our central schoolhouse, though. Don Irwin, and Leo Jacobs, the project architect, were able to convince Colonel Hunt that one central school was better than four smaller schools scattered around the valley.

They showed Hunt the materials and equipment on hand specifically for a central school building and reminded him that buses already in use around the valley were for the school. They also reminded him that without the central school, there would be no building in town large enough for public gatherings.

Even if Hunt preferred the local-school concept, he saw the logic of not wasting money already spent. He agreed, and work was started on the central school building.[4]

Under Roof

While children turned their attention to school, the main focus for the rest of us during September was completing houses. Carpenters were finally working on our house, and since Neil was at school all day, I became the building "boss." You can imagine how thrilled the carpenters must have been over having a woman hanging around.

The Miller's home northwest of Palmer. The house, which has gone through many additions and modifications, can still be seen from Fishhook Road.

We decided to really personalize our house and construct a flat roof. Construction officials warned against it because of the roofing materials being used, and the carpenters thought the idea was pretty screwy, but we persisted. They finally allowed us to build our flat roof, and in newspaper pictures our house was referred to as having "Spanish styling."

By early September our family was the only one in camp still living in a tent. The Baileys, who we dearly loved, found their land to be worthless. They traded their tract for one nearer to Palmer, so had to move to a different camp. The McCormicks left the colony entirely. Mr. McCormick wanted to stay, and tried so hard to get his tract fixed up. Their house was almost finished, with even the well and pump in. Mrs. McCormick was determined to leave, however, and she nagged and howled at him until he just gave up. She never even really tried to make a go of it here. Evidently, their marriage had always been a rocky one, and the relief agency sent them to Alaska to try and patch it up. Mr. McCormick told some of the men that he would never return to the States with her, and planned to stay in Alaska somewhere, so I don't know what happened to them.

The rest of our neighbors had moved to their tracts. Since there was no well on our tract, we decided to stay in the tent as long as possible. Otherwise all Neil's time at home would have been spent hauling water.

By living in the tent Neil had a few hours each night to work on the house. Only one side of our house had the studs and sheeting up. Another side was ready to put the sheeting on. I had never seen houses put up like they did those. Usually I saw them put up the studding then nail the sheeting on. But, in some instances, these even have the building paper, siding, and paint on before they raised the wall.

That summer and fall provided the girls and me with quite an architectural education. We got so we knew something of what it was all about; what words such as joists, studs, and sheeting meant. We could actually recognize different pieces of lumber by name, and we helped nail on the sub-flooring. Neil cut and placed the ship-lap and we drove the nails to fasten them to the joists. We felt it was quite an achievement. How much more fun it was to live in a house we actually helped to build, even though it was only a little now and then.

By mid-September our house was framed and roofed, and we eagerly anticipated its completion. The house was just far enough along so I could sort of imagine what it was going to be like. The rooms were smaller than they seemed in the plan, but they were plenty large enough, and so compact and convenient. I could hardly wait to get moved in.

Under Roof

We endured life in the tent while waiting for the house to be finished. Being home all day, I suffered most from September's chill and dampness. I learned later that frosts always came early to our part of the valley, and of course, the houses in our part of the valley were some of those finished last. In September some places did not thaw out all day, but, the sun was lovely. We who were still in tents felt the cold, especially nights, or almost any time when inactive. I caught a cold, and when I thought it was about cleared up I did the washing. After that it took me a week to feel warm again.

The chilly tent did not restrict our social life, however. Colonel Hunt, and Ross Sheely, the assistant project manager, came to dinner September 2nd all dressed up in "conventional blue serge" with a box of very fine chocolates for their hostess. (I believe it was the first time Hunt wore civilian clothes since he arrived in Alaska.) The only kind of meat I could get was a very poor cut of pork, and it didn't brown as it should, but anyway that was their fault—I was supposed to have moose! They had gone hunting and promised to provide a roast, but they failed to bag their moose.

The Colonel and Ross entertained us on the piano, playing 'Chopsticks' and 'Once I Had a Charming Beau' very expertly with variations and trills and additions I'd never heard before, also a remembered music lesson or two, and a little chording. But our family enjoyed it anyway and they seemed to enjoy doing it. We played anagrams and dominoes, and the girls had the men write in their autograph albums. Anyway, it was our first real attempt at entertaining in Alaska—and done in a 16'x24' tent.

I did receive some wild game from their later hunting trips. Many of the colonists took advantage of the abundant wild game in the valley that fall. In fact, some appeared to be more interested in hunting than farming. We had to live in the territory one year before being classified as residents, and the Alaska Game Commission affirmed territorial regulations that we had to have non-resident hunting licenses. Although few of us could afford the $50 non-resident license, quite a few moose and bear were shot that fall by colonists—none by us, though. There was enough illegal hunting to warrant the appointment of a deputy game warden for the Matanuska Valley in October.[5]

In addition to hunting, fall also brought school dances and other social events. Mardie began dating that fall. After supper one evening, the two Lamp boys (settler neighbors in Mardie's class) came with a note from Mrs. Cornelius (another settler) that if I would let Mardie go to a party at Wasilla that night with the Lamp boys, who she would vouch for being OK., she would let her daughter Ruthie go. Ruthie was also in Mardie's class. Mardie was quite thrilled, though she tried hard to act as though it didn't mean much to her.

That was her first real experience going places "coupled off," but the settler children were such wholesome folks I did not worry much. Mardie said she and Ruthie drew straws to see which boy they rode with. The boys drove a Ford truck. Ruthie and Leonard rode in front going over, and Mardie and Jerry in the back. On the way home Mardie and Jerry drove, and Ruth and Leonard rode in back. They had tire trouble and before they finally got Mardie home they had to go home and get their dad's big car. So you see, Alaska was quite like the states after all.

We were still living in the tent by mid-September, but hoped to be out any day. The house exterior was complete except for painting, and Neil and the carpenters worked furiously to get the interior complete enough to move into. We were anxious to move in, but progress seemed slow. We wanted to finish the floors and walls in the kitchen and adjoining room before moving in. Then we could live in those rooms while finishing the rest of the house. Consequently, we shivered in the tent a few more days.

Furniture orders began arriving in late September. In keeping with everything else happening that summer, there were many mix-ups and duplicate orders. Some colonists' orders were missing, and other received substitutions. A few people didn't like their furniture when it arrived and turned it back in. With the shortage of warehouse space, it was quite a situation. I received three boilers along with my gas-powered washing machine. The day our washing machine was delivered was my laundry day, and after finishing 10 tubs of clothes by hand at the tent I walked up to the house to check on its progress. There in the front yard I discovered the new washing machine, ready to be used.

Although the ARRC provided transportation for the basic functions of the project, we, like other colonists, had other needs and activities that required transportation. Corporation trucks delivered building supplies, groceries, livestock feed, and other goods to the camps, and buses took children to school and men to work during the week, but for most other non-essential activities, we caught rides with delivery trucks or with others when the opportunity arose, or we walked. Few people had private automobiles. Neil and I often walked the five miles to Palmer or even to Matanuska five miles beyond.

By the end of September I was very anxious to move into the house, so I walked to Palmer, even though I had a cold, to pick up a heater to dry out the house interior. I walked part of the way home and was all in when I got there, with my cold worse than ever.

I felt I had get into the house to get rid of my cold, so Neil and I worked in the cold over at the house getting celotex and plyboard up to enclose the kitchen and small room adjoining. I coughed all night and the next day Neil made me stay in bed. I did not feel I was ill enough to warrant that, but anyway I stayed in bed, and was cold! The wind blows hard nearly constantly at that time of year, and the tent was almost impossible to heat. So we decided the time had come to move.

September 26, the day of the move, was a school day. We had hoped to put skids on the tent and have it, along with all our possessions, hauled by tractor to the house. The tent proved too wide to navigate the road to our tract, though, and we had to move everything piecemeal. Since Neil taught school all day, several neighbor men offered to help us move. It did not take long for them to move our possessions into the house, but the packing and moving was not exactly an orderly process. It took me quite a while to sort through everything once it was moved. But we were so thrilled to be in the house. It seemed too good to be true! All that first day I enjoyed just looking about me with the satisfaction of feeling that "this was ours!"

Of course, the place did not look the same in our eyes as it would in any others, because when we saw it, it was touched up with the plans we had for it in the future. To the casual observers it was merely a newly finished exterior, with a flat roof and shavings, boards, nails, and confusion all around. They could not see the lovely sloping lawn and the shrubs that we could. Inside, we saw it as we hoped to have it finished, furnished with furniture of our own design and make. To outsiders the windows were not even washed, the partitions were not all in, and the celotex and plywood that was up was what was temporarily boxing off two rooms for us to go on camping in.

The same day we moved into our house, the *Valley Pioneer* reported that construction of the last house in the project (belonging to Lloyd Bell) had been started. The article went on to say 101 houses were roofed, and all others (73) were in various stages of construction. Fifteen barns were completed and the bakery and cobbler shop erected. The hospital's interior was completed and the power plant generators were successfully tested. Bert's Drugstore, operated by Bert Weeda, opened in Palmer, as did Sally's Sweet Shop, and a soda fountain adjacent to Koslosky's Department store.

Since most of us were at least "under roof," well drilling delays became the next major concern. The *Valley Pioneer* reported that, "So many people jump Ross Sheely about wells every day he has developed a severe case of 'artesianitis'. This is an odd malady where the victim dreams continuously of driving a pipe into the ground three or four feet and then getting it blown out of his hand by a gusher of artesian water.

Under Roof

"Ross was right in the middle of one of these spells when an idea struck him. He is going to buy an old abandoned railroad tunnel up here on the deserted Chickaloon branch, split it up into well sizes and distribute them free. Get your order in early, Ross is likely to wake up out of his dream any time."[6]

Many people in our part of the valley felt that Palmer was too far to go for church, so in September, our family helped organize a Sunday School for that part of the valley. The service was held in the afternoon at the home of the Brazils, a settler neighbor. Our family and two children from Camp 5 were the only colonists the first Sunday, but we had a very nice turnout of settlers.

Adam and Fannie Werner's homestead along Fishhook Road near the Millers. Adam began the homestead in 1914.

I liked the settlers. (We called all who came here before we did "settlers" just to distinguish them from "colonists.") They were such an earnest, clear-headed group.

They were folks who came in here with only a few dollars and cut their own way without help from the government like we were getting, and they were making it pay. I liked their attitude . They were not sitting around waiting for somebody to hand them something on a silver tray. They were out working and planning and sacrificing and making things go. It must have amused them to listen to some of the colonists.

I wish it was possible for everyone to have seen one of the settlers' farms. The settler had built up a place, all of hand-hewn logs, that was truly a show place. Everything he had to bring in he carried from the little town of Knik (about a 30 mile hike). His eight-room house was the most beautiful logging job I had seen in the valley. Every log seemed the same size as the other—all so carefully hewn as to look like sawn lumber. (Many "doubting Thomas's" were taken in there by Colonel Hunt to show them what could be done if one had the desire to make the effort.) One could hardly class me as a "doubting Thomas" -- but seeing that place confirmed my own opinions of the opportunity that lies here.

As you drove in the place you went through the loveliest heavy stand of oats and peas, and the first glimpse of the place was breathtaking. There was a lawn—an actually mowed lawn! There were beds of huge yellow pansies, yellow and white California poppies, shrubs, and hollyhocks, all tastefully grouped around the house, and vines in baskets hanging on the porch.

The settler's wife loved to have visitors and was so proud of her home and all they had accomplished. They had an oil drum heater in the cellar, no hot-air pipes, just radiators in the floors and the heat rises. She said the floors were so warm all winter that she and the two children went barefoot all winter.

She showed me the newly-floored chicken house. They ran a heater in the chicken house all winter, and shipped in all the feed from the States, and still found that their eggs were one of their best paying crops. We went through her berry patch and I ate raspberries as big as the end of my thumb, just the largest juiciest raspberries one could ever dream about.

She took me in the back room and showed me the 85-pounds of cheese she had just finished making. They were put up in wheels and parafinned. It looked like regular American cheese.

Bachelor settler's tract east of Palmer.

Of course, not all the settlers were as ambitious, especially some of the bachelors. Some of their homesteads were more winter retreats than full-time enterprises—where they could hole up when not hunting, trapping, or prospecting.

During the colony's first year, the ARRC had a rather strained relationship with valley settlers. The settlers did benefit from the improved roads and the school and other community facilities the Corporation provided. However, while many settlers befriended and shared their hard-won expertise with colonists, some settlers bemoaned that they, in contrast to colonists, had moved to the valley without government assistance, had established their farmers without government financing, and, aside from assistance from the federally-funded Matanuska Experiment Station, had no access to government programs and government-provided equipment. They also resented that colonists could buy supplies from the Corporation on credit at prices the settlers viewed as government-subsidized, while settlers could not buy from the Corporation at all, even with cash.

A Creek, a Hill, & a Forty

It seems, though, that the ARRC came to believe that if the project was to succeed, both colonists and settlers needed to be united in a valley-wide marketing cooperative. Consequently, the Corporation (at the behest of Colonesl Westbrook in Washington, D.C.), began making loans to settlers for improvements to their farms, as long as the settlers agreed to join the farmers cooperating association (which had not yet been organized). Settlers were further integrated into the project through actions such as purchasing privileges at the commissary, jobs provided by the Corporation (later the MVFCA), as well as participation in the co-op's profit-sharing program.[7]

I was impressed by the settlers' fields and gardens, but was disappointed in Camp 7's community garden in August. However, by September the community garden had improved, and we began harvesting the cabbages, rutabagas, carrots, peas, head lettuce, and celery from our part of the garden. In addition to approximately 100 acres in the project's community gardens, 35 acres were planted in carrots, potatoes, turnips, and rutabagas.

Oat-pea hay was raised on the remaining 40 acres of the project's cleared land. A combination of oats and field peas, the oat-pea hay produced in the valley was reported by the Matanuska Experimental Farm to be superior to hay produced in many prime farming areas of the United States. An additional 125 acres of cleared land were rented from settlers for hay production.[8]

By late September our Sunday School was still quite a thriving concern. That was no mean feat, considering that it was harvest time, and everybody, children too, were helping in the fields on all sunshiny days—Sundays included. Our Sunday School still attracted about 12 families, and that included only this neighborhood. Similar Sunday Schools were being held in the other neighborhoods that were also too far out to reach Palmer comfortably.

We would work on the house in the morning, teach Sunday School in the first part of the afternoon, and come back to work on the house again in the evening. If it was a nice day we would work outside. When the sun shown, everybody worked on Sundays during the busy times.

The life of our community church was tangled in the fall, when several of the religious groups in Palmer refused to be united in a single community church. As a consequence, the Corporation announced it wasn't going to build the non-denominational church as originally proposed, and it would not provide Corporation land for any of the churches. Each religious group began plans for constructing their own church, but the Presbyterian Church, which paid Reverend Bingle's salary, balked at providing us with construction funds. They had heard such adverse publicity about the project, they were afraid it would fail. We had a church meeting, and agreed to try and build a temporary building to house the minister's family and provide room for services.[9]

The reactions resulting from all the adverse publicity seemed too bad. The families that caused most of the problems, and did most of the howling, had already returned to the States, but the rest of us had to suffer for their beefing. The activities of late summer and early fall kept most of us colonists busy, though, probably reducing some colonists' feelings of discontent. Five families left the project in August, but only one left in September. Winter was coming and we were settling in.

Seven

This is fun

I was extremely pleased that life was settling down after the trying summer and fall, and that Neil and the rest of the family were enjoying themselves. Neil was so happy being busy around the house. His grin was permanently back, I believe. He was up early in the morning and late at night, building this, fixing that, hammering here and sawing there, all the time fixing something for his family's convenience and happiness. I tried to get him to lay off for a few nights and get some rest, but he claimed he didn't get tired doing those things—they were fun.

We were having the time of our lives! It was so good to have Neil grinning over things. Life may have seemed hectic and hurried, but it was good. I was also very grateful to be out of the tent and into my own house. Not all colonists were as fortunate, though. Some waited until early November to move into their homes. The cold and rains of October and November were bad enough if you were in a house—life was almost intolerable in a tent.

The valley received almost twice it's normal precipitation of 2.6 inches during October and November, 1935. Some roads were so muddy that tractor-drawn sledges were used to transport freight and people.[1]

The bad road conditions made life miserable for almost everyone. Some families, like the Henry Rossiters, experienced more than their fair share of misery. Twin girls were born to the Rossiters on September 29.[2] The day arrived for Mrs. Rossiter to leave the hospital and her husband made big plans for the homecoming. The furniture for their house had arrived and he got a truck to get it hauled out the day she and twins were to get home. While he moved furniture, there were three or four other small children at home waiting. The doctor arranged with Colonel

This is Fun

Mary Rossiter and her twin daughters.

Hunt to take Mrs. Rossiter and twins out after lunch. The weather was vile—rainy and cold—and the roads out there were rough and slippery, so Ross Sheely went along in case he was needed. Just before they reached the house they came to the truck-load of furniture mired in the middle of the road, with Mr. Rossiter plastered with mud trying to get the truck going.

Hunt's car couldn't get by. So the nurse who had gone along to get them established at home picked up the clothes basket with one twin, Ross took the other basket and twin, the Colonel carried Mrs. Rossiter, and they sloshed up to the house.

They got inside and the house was bare except for a pile of bedding on the floor where they deposited the mother. The children were dancing around so glad to have "mama" home, but the fire was dead and everything was cold. You could just feel how disappointed Mr. Rossiter must have been. Here he had planned to have everything all so cozy and right, and it turned out like that. Some way I feel sorrier for him in his disappointment than Mrs. Rossiter, who must surely have been disappointed too.

There had been a few babies brought into the world at Palmer during the summer, but a regular bulge in the birth rate was observed that fall, curiously occurring about nine months after most colonists received word they were moving to Alaska. I think that by that fall about 75% of the women in the valley had either already had babies or were expecting.

While rain bogged down vehicles and made walks to town almost impossible, we did appreciate the rain water since we still had no well. We preferred it to creek water for drinking, and I used it when I could for washing. We were slowly finishing the interior of our house rather than paying a carpenter the $150 to $200 it would cost to have it finished for us.

Our house was spartanly furnished, but we didn't mind too much. I did occasionally looked longingly at my neighbors' furnished houses. I have to admit, I was a bit envious, and had to keep telling myself that it was better not to go into debt when Neil would eventually get ours done. It was quite a strong argument that at least we would have individuality in furniture that Neil made, instead of getting it from Sears and Wards like most of the colonists did. But I got impatient waiting when all about us were getting so nicely settled. Someway the army cots and salmon kegs seemed awfully "hard sitting" after returning from our neighbors' davenports and easy chairs!

We furnished our house with the beds and home-made tables and chairs from the tent, and trunks, boxes, and cases of food. Our piano was the only "real furniture" in the living room. Tables consisted of barrels of unpacked dishes and other household goods. Chairs were the girls' chests, a trunk and a keg of salt fish. The davenport was the winter supply of case goods, covered with four spare mattresses and an army blanket.

The only new furnishings were the heater we bought in September to dry the house interior, and a new kitchen range bought in October to replace the flimsy stove used in the tent. The Corporation received so many complaints about the stoves that it sent two of them back to the firm they were purchased from protesting the price and quality.[3]

We did not let the lack of proper furniture stop us from having guests over, but our main activity in the evenings was finishing the house. In October we managed to get the floor laid in the kitchen and storage room, finished the cellar steps, closed up the cellar and banked the house.

By the end of October the major goal for 1935 had been achieved. Almost all houses were completed and hopefully all colonists would soon be moved into their homes. The same could not be said for project officials and workers, though. Carpenters continued to work on staff cottages and a dormitory, as well as erecting five 20-man barracks for temporary workers. Railroad boxcars were also moved on to sidings in Palmer and converted into bunkhouses for workers. Even so, many workers and officials continued living in tents (now insulated) deep into winter.[4]

Land clearing had ended for the season, and house construction was tapering off, so the ARRC began sending transient workers back to California. A small number of skilled transients were retained to work on the buildings in the civic center, to drill wells, and do other jobs. According to the *Valley Pioneer,* over 200 transients applied for the approximately 30 positions available.[5] The Corporation also began advertising for Alaska residents to fill additional skilled construction jobs.

With the construction season ending, most of the buses were no longer needed, so two of them were turned over to the Territory on October 20. A third bus was retained by the Corporation. The transfer meant that Neil would now drive a school bus. It also meant we wouldn't be moving to Wasilla.

Driving a bus provided Neil with an additional $40 per month, but first he had to get his bus drivable. They were in bad shape from constant usage during summer and fall, and could not be totally fixed until new parts arrived. After some major work, Neil's bus would run, but it had a bad battery and had to be crank started. At night he parked the bus up on the hill by the house so he could run it down the hill in the morning to start it.

I hated to see the family start out for a 13-mile ride, even in a heated bus. There was one strip of road of three of four miles that was a mean, narrow, winding, slippery thing, and by then

Neil and his bus.

there was a light covering of snow making it that much "meaner." There was just enough snow to make it slippery, but not enough to fill in the ruts.

Bus driving was not pleasant during winter, but at least Neil and the girls stayed inside most of the day once they arrived at Wasilla. I still had to do chores at home. Washing clothes could be done inside, but the clothes had to be hung to dry outside. On one occasion I was washing and having difficulty keeping the clothes on the line, and I noticed the tarp covering the cow blowing off. We had been keeping the cow under a tarp behind the hay pile, and Neil left her quite snugly closed in with blankets over the door. Most of the wind came from the east, and she was back of the hay to the west, but that day a strong wind started up from the west carrying hard pellets of snow before it.

I went down and thought I fixed the tarp securely, but when I had finished climbing the hill back to the house I heard a swish and there was the tarp gone off the cow, hay, frame, and everything. So I took the cow over and put her in Johnson's barn. I came back and covered the hay and started up my washing again when I saw the tarp that covered the lumber sailing across space and I had to go replace that.

This is Fun

The Corporation stopped providing general transportation for colonists when it transferred the buses to the Territory, so we purchased a "new-used" Ford sedan in October to fill the void. It was not exactly dependable transportation and we nicknamed it the "Derelict." The same day the tarp blew off the cow, the Johnsons borrowed the car to drive to Palmer for supplies, and I went along. We got started about four o'clock, and about half way to town, the Ford stalled and refused to function. So we walked back home.

In the meantime, Neil and the girls came home. They didn't realize the wind had blown so hard in the morning, but they saw the barn was down, the outhouse door was ripped from its hinges, the car was gone, and there were tracks all over the yard. (Bagsy had come through our place with the horses dragging some logs.) Neil and the girls assumed someone must have been hurt and rushed to the hospital in the Ford. So they spent some anxious moments until we all trudged up the road.

On October 25, what the *Valley Pioneer* described as the "social event of the year' occurred.[6] The project's public health nurse, Madelon de Foras, married Eugene Ellsworth Sedille, a construction official. The ceremony was at the officers' mess at the construction camp, and a reception, luncheon, and dance were held afterwards. It came out during the ceremony that de Foras was a French countessa, which impressed everyone greatly. The de Foras-Sedille marriage was supposed to be the first in the colony, but another couple, learning the wedding date, beat them out by a week.

Everyone in the valley was invited and the ceremony was well attended. I wanted to go but it would have meant walking. The construction camp was a mile beyond Palmer, the temperature was hovering around zero, and the wind was blowing, so I decided to stay home. Fortunately, Colonel Hunt, attired in his dress uniform, with all its bars, buttons, shiney leather, and Croix de Guerre, stopped by afterwards to fill us in on the details. He gave the bride away, and although he felt uncomfortable in his dress uniform at a civilian wedding, the new Mrs. Sedille had insisted.

In November, the Corporation did start providing Sunday bus service to Palmer for church services, charged to general project welfare. Although the Department of Education now owned the buses, the Corporation rented them for social events.

Since Neil drove the bus on weekdays, it also fell to him to drive it on Sundays. We were regular church goers, so the arrangement worked out fine.

The Corporation still provided transportation to committee meetings in Palmer, and on one trip to the council meeting in October, the driver of the pick-up (a transient) asked me how old my eldest daughter was. I told him that Mardie was almost 14, and he replied, "Ah! Old enough to date! Do you let her date?"

Mardie hadn't shown any leaning towards dating yet, and I informed him that she still played with paper dolls, but he said he was going to stay in Alaska at least until she would date him. (He was about 25 years old.)

He had had a drink or two I think, and he raved about her "ethereal beauty" and so on. He said he was determined, so he'd be "initiative" and show a little "stragedy" to the mother first.

I wasn't sure what he meant by that, but several days later he was the driver for the supply truck, and he came in to get the requisition signed. I had just mopped the floor, had cleaned up, and was putting fresh cookies on the table when he came in.

He announced to his two co-workers, "Here's a real housekeeper, guys—look at them floors!" They all raved about my housekeeping, so I had to pay back their compliments by giving them cookies.

Then the driver said, "Ha! I knew I was picking the right mother-in-law." I suppose that whole episode demonstrated his "initiative" and "stragedy", but I used a little "stragedy" myself and got them to carry 22 gallons of water from the creek a 1/2 mile away. So who came out ahead?

This is Fun

On November 1st the last large group of transient workers left for California. The skilled labor hired in Alaska, and 47 skilled transient workers, remained to complete construction.[7]

Colonel Leroy Hunt, who replaced Mr. Fuller as FERA representative

Colonel Hunt also left Alaska the same day. We hated to say goodbye to him, like we hated to say goodbye to friends when we came up here. When Mr. Fuller left Hunt in charge here it was one of the best things to have happened to the colony. When he chose a man from the service it was the first wise move, since he would have no political ax to grind. But his choice of a man of Hunt's personality was of even greater value.

Hunt's philosophy of life was that there was no greater happiness to be found than that which came from doing something to help another. We all benefited by that. By his diplomatic and kindly handling of the executive staff, as well as the colony and transient personnel, he, in his quiet way, did more to bring about the speedy construction results than any other man in the entire set up.

He was also responsible for the feeling of good fellowship and general happiness that prevailed throughout the colony. Our family certainly missed him stopping in for tea, and having the heel of the loaf on baking day. We felt he was part of the family. Of course, about 100 other families in the valley probably felt the same way.

Before leaving Palmer, Hunt oversaw several changes in the project. The ARRC was reorganized at a board of directors meeting October 29 and 30, and several board members were replaced. Colonel Ohlsen of the Alaska Railroad was elected president, and Luther Hess, a lawyer from Fairbanks, was elected vice-president. Ross Sheely was appointed to the board, and William Bouwens, a colonist and president of the community council, was also appointed. The Corporation headquarters were transferred from Juneau to Palmer. With it's reorganization, the ARRC took over active direction of the colony, although it still reported activities to Colonel Westbrook and FERA in Washington.[8]

Hunt appointed Ross Sheely as general manager of the project, replacing Don Irwin. Don said that the decision to replace him by Ross was actually made back in mid-August. Washington officials felt that Don lacked the ability to definitely say yes or no—being too kind hearted. Ross, on the other hand, had the push, the drive, and the ability to make decisions and be final.

Don was ready to quit the project entirely and return to the Experimental Station, but his expertise was too valuable to lose, and Hunt offered him the job of assistant manager. Don remained unswayed, but the Colonel finally convinced him to stay by showing him a petition circulated among the colonists stating that if he left, they would quit the project. He was well liked by the colonists, and his demotion was not popular. Although Don surely must have been stung by the turn of events, he did not let it stand in the way of friendship and professionalism.

Don and Ross were old friends of long standing and after an initial adjustment period, they worked very well together. Don told me that the reasons for his demotion were never adequately explained to him. I felt guilty—listening to his hurt—feeling the hurt with him—and knowing why and still not telling him. I knew how well Colonel Hunt liked Don and how he hated to have to make the change, because Hunt has used me as a sounding board for most of one afternoon while he tried to make up his mind. although I liked Hunt very much, I feel he slipped up on his obligations by not explaining things fully to Don.

Ross Sheely wasted no time after his appointment in moving his family to Palmer. For the previous six months he had been "baching" it while his wife and three children remained in Fairbanks.

When the Sheelys left Fairbanks at the end of October, it was 37 degrees below zero, quite a difference from our own "balmy" weather. Mrs. Sheely was one of the most interesting women I had met in Alaska, and their children, ages 15, 13, and 10, worked in quite nicely with ours.

On November 4th the last colonist in a tent moved into his house. The housing situation was helped by the return of some colonists to the states. Of the 203 families in the project at the beginning of the summer, 38 had returned, and only 165 houses were required in 1935.[9]

During October and November, house warmings were the main social attraction. They were so numerous on one weekend that an American Legion dinner had to be postponed. There were also many birthday parties, wedding anniversaries, and baby showers. The ARRC sponsored a children's pet show (one of the winners was a bear cub); there was an open house for the new community hall near Bodenburg Butte ; and a turkey shoot was held near Wasilla Lake.[10]

As soon as the new hall at Bodenburg Butte was completed, it was pressed in to service as a schoolhouse for children in the area, relieving teachers of the duty to visit students at home. The Alaska Railroad also brought a passenger car into Palmer in November to use as a schoolhouse.

Reverend Bingle hosted a church social November 23, celebrating the completion of our new church/manse. Our church had acquired a lot in town after it became apparent that a community church would not be constructed, and volunteers from the congregation erected a building during October and November. It was just a house, but with a 16 foot by 24 foot living room for services. Because of Neil's job, we weren't able to work on the manse, but we were able to contribute financially. Our congregation's next challenge was to find a building site and erect a permanent church building.

The colony council kept me active. An election board was appointed at the November 11 meeting to prepare for election of new representatives to council. The new council was to have a slightly different make-up. The valley was divided into four instead of nine districts, and two representatives would be elected from each district: one colonist and one settler.

I was on the election board and had to spend an entire day preparing lists of families in each district to be sent to every household in the area. A sort of primary was held first. Candidates were nominated from the lists, and the three names receiving the highest number of nominations in each district would be the finalists in the election. The final voting would occur in December and the new council members would take office January 1.

An American Legion post was organized at Palmer in October and Neil became very active in its activities. Elections were held November 4 and Neil was elected post commander. Regular Legion meetings were held at the Bingles. Having Neil gone was not so bad, but the Ford had burned out a bearing. Neil didn't have time to fix it, so he walked to the meetings. With the temperature hovering around zero degrees or lower, and the wind blowing, I did not envy him his walk. But he felt he had to go as long as he was commander.

The Anchorage Legion and Auxiliary came down November 28 to install the Palmer Legion officers. The installation, and the dance following it, were held at Matanuska, where there was a large community hall. Bus service was provided for the evening, and the dance was open to the public, so there was a real crowd.

Taking the bus from Camp 7, we had to ride about 20 miles through Camps 2 and 4 picking up passengers before arriving at Matanuska. The bus had a very poor heater, the night was freezing, and the roads slippery. We were all nearly frozen when we finally arrived. By the time we reached the hall, the Anchorage Post and Auxiliary were already there.

The ride home wasn't any better. We had to wait until 2 a.m. before starting home, and half way to Palmer our bus broke down. Some people caught rides home, but others, including us, had to go back to Matanuska and wait for the second bus. We finally headed home at 5 a.m.

The bus's maximum load was 32 people, but 45 crowded into that last trip. First we had to drop people off at camps 8 and 9, and then drive into Palmer. We had borrowed dishes from the mess hall, and had to return and wash them before breakfast. Gus Rascke from our camp was the bus driver, and he was unfamiliar with some of the roads. The windows were heavily coated with frost—there was only a small peep hole thawed out in the windshield—but we finally made it to Palmer after quite a few missed turns.

After stopping at the mess hall, we continued on with people from camps 5, 6, and 7. Gus was so worn out that Neil eventually took the wheel, and we finally got home at 7:15 a.m. Everyone was frozen stiff, and the fires were out, but we made coffee and had breakfast. Gus was still at our place when a truck arrived to fetch him so he could haul hay. Then Neil had to start right back with the bus to pick up the church load. He took his load in, then started on the other route (because he had their bus) until he reached the other driver's home, where he turned over the bus and its occupants. Then he hiked down to Matanuska, fixed his own bus, picked up his church load from town, and got home at 2:30 p.m.

I was hauling water when he arrived and he helped finish, so Neil didn't get to bed until 4:00. He slept till 7:00, had supper, worked on the woodpile, and then on the bus's spark plugs until 10:30. It's no wonder he felt relieved to get back to the calm of the school room on Monday morning.

In addition to the installation and dance, the Palmer Legion Post sponsored a father-son banquet and several other dances during October and November. Naturally, since Neil was active in the Legion, I became active in the Auxiliary.

While there were welcome respites from daily chores and work, the problems of building a new community soon returned. All colonists were housed, but it became apparent that not all of us would get barns that winter. Officials knew arrangements had to be made to stable all animals. The valley was canvassed for available space, and colonists' animals were stabled in adjacent colonists' and settlers' barns, at the University Experiment Farm, and in other available buildings.[11] Arabella, our cow, stayed at the Johnson's farm.

Since Neil and the girls were gone from about 6 a.m. to 5 p.m. on weekdays it fell to me to feed and water Arabella. I had to haul water from a spring about a quarter of a mile from Johnson's. When I got to the spring I had to climb down the slippery bank, break the ice over the boxed-in place where we dipped our pails, get two pails of ice and water, carry it through the brush to the Johnson's, put it in a tub, and mix it with warm tea kettle water that I carried from the house. I let Arabella drink that, then I would carry two more pails and leave that in the tub to heat in the barn for her to drink during the day.

The water froze to the sides and bottom of pails just between the spring and the barn. We carried the hay, rolled into a canvas bundle, from here over there.

For a while my shoulder muscles were so stiff from carrying that I could hardly move my arms without being conscious of it. But they say the first hundred years are the hardest, and eventually those muscles hardened to it!

Four of us in camp 7—the four in the middle of the camp, and the four farthest from the creek—had no wells. Each week we were promised they would surely be here, and each week it was put off again. In the meantime, we received a taste of "pioneering." Before, we were pioneering "deluxe."

The valley was having a typical winter and by mid-November rain was still alternating with snow. [Temperatures fluctuated between -5 degrees and +62 degrees Fahrenheit in October and November [12], so roads were often hazardous.]

It would snow, then rain and thaw and freeze again, which played general havoc with the low places in the roads. The streams were over full, then ice jammed and froze, and the streams hunted new channels. As a consequence the water was from one to three feet deep in some places on the roads. No damage except to cripple transportation.

We had one bad place right here in our neighborhood. If we walked anywhere, we had to wade through that spot in freezing water up to our knees. Yesterday Neil got stuck there with his bus. He finally got out but his brakes were frozen, so he spent most of the rest of the day fixing them and did not get to school at all. The rest of the time he worked with ax and shovel breaking up the ice jam that was causing the difficulty. He also had to break up the ice about three miles along his bus route where there was another bad place.

At least Neil got a new battery for the bus in November and didn't have to crank start it. He only had to run it before going to bed, and then again about one or two in the morning to make sure it started after breakfast. He was also able to get the old tent and frame converted into a makeshift garage, which protected the bus from the brunt of the weather when Neil had to work on it, which was all too often.

By Thanksgiving we had finished the living room floor and had the linoleum on the kitchen floor. The walls and ceiling were insulated, and Neil was putting interior partitions and plywood up in the cellar and storage room.

That was the first Thanksgiving I missed some sort of family reunion—if not our family's, then someone else's. Thanksgiving wasn't very "holidayish" here with us. Tuesday night Neil's bus slipped off the road on the way home from school. He spent all day Wednesday in town or at the bus. I had spent Tuesday in town and stayed in for council meeting in the evening and got home at three in the morning on Wednesday. Then I had to go to town Wednesday evening to count votes. So I didn't get really festively prepared, though the girls and I did hang the bedroom curtains that we had in Blair at the living room windows. We put my kitchen curtains up, also temporarily. Then just before supper last night Janell and I gathered some spruce boughs with cones and made a table piece of it and had tall yellow candles.

We had our first taste of cold storage chicken shipped in from the States, with potatoes, carrots, rutabagas and cranberries raised in Alaska, and pickles and pumpkin pie. We put on a table cloth (unusual here) and all the food dishes and matched silver and ate too much, trying to make it seem like Thanksgiving. We were of course thankful for our new start in Alaska, and while the project's construction was slower than anticipated, it was comforting to see its progress. Palmer was gradually growing out of tents into buildings. The general store, drug store, and sweet shop were not part of the Corporation, just branches of houses in Anchorage. The general store was much like most small town general stores but the drug store and sweet shop were as good as in towns of pretty fair size, and they did a good business.

Our Civic Center, across the tracks from the "tent city," included the new Corporation buildings that are now beginning to be occupied and functioning normally. The commissary and most of the offices, the new warehouse, the barber shop, the cobbler shop and the women's exchange and the garage were all completed and in use. The permanent hospital building was almost finished and the school, the community hall, the dormitories, the manager's, assistant manager's and doctor's houses were well along. They seemed to be framed on the outside anyway.

Eight
Matanuska Valley Christmas

Winter view of the Miller farm in mid-1930s

With Thanksgiving over, everyone set their sights on Christmas. But, while we and other families in the valley began preparing for Christmas, a tragedy occurred that dampened spirits considerably.

The house of George Emberg (a colonist) burned to the ground the morning of November 30, and his wife and daughter were gravely injured.[1]

George, who had a side business providing milk to people in his camp, was anxious to start deliveries, and the house caught fire when he threw what he supposed was kerosene in the kitchen stove to quicken the fire. It was gasoline instead, and flames erupted from the stove. The partly-filled gasoline can exploded and the house was quickly engulfed in flames.

Everyone escaped from the house, but Mrs. Emberg and their three-year old daughter, Dixie, received second and third degree burns over most of their bodies. Mr. Emberg and his brother received several major burns, and two other children escaped uninjured.

The Embergs were rushed to the Palmer hospital, and blood transfusions were started on Mrs. Emberg and Dixie, but there was no real hope for Mrs. Emberg and she died the next day. Dixie was a real fighter though, and the entire community prayed for her recovery, contributing funds and even starting plans to send her to a plastic surgeon. The ordeal was too much, though, and she died three days later. According to the *Valley Pioneer*, she was buried next to her mother in the Palmer cemetery, with the children from her camp bearing her coffin the final few feet to her grave.[2]

The valley's entire population was shocked by the fire. The only consolation in the tragedy was that the fire was not caused by faulty construction, or something else that might have been blamed on the Corporation. For with the papers and magazines so anxious to get their teeth into something they could use to belittle the project, how they would have gloried in anything like that.

We found out in early December that I was running for the community council seat in our district. I had not indicated that I wanted to run, but enough people had nominated me so that I qualified for the final voting. I enjoyed being on the council, but was not really into politics and did not campaign for re-election. The Corporation also made the decision to allow settlers to be included in the council.[3]

Christmas plans for the valley were in full swing by early December. The big celebration in Wasilla would be the program at the school. Naturally, Neil and the girls were involved in it.

Social committees in Palmer had met several times already to plan for a community celebration at the Palmer community hall, which the Corporation promised to have completed by Christmas. The celebration was for both colonists and settlers. The American Legion was responsible for putting up the Christmas tree and decorating the hall, and the school teachers were to coordinate the program. The ARRC would rent school buses for transportation.

No one could feel very sorry for any of the children in the valley. The Corporation received a telegram from Sears and Roebuck stating that the company was sending gifts for all children up to 14 years old, and the Junior Red Cross was also sending gifts. The Alaska Steamship Line notified project officials that the company's Christmas show featuring Santa Claus and entertainers would stop in Palmer on it's annual trip to Alaska. Funds were being donated by local workers and residents for candy, nuts, and fruit to be distributed at the community party. Even if the children had no more Christmas than that, they surely would have been happy enough.

Jack Allman wrote in the *Valley Pioneer* that the Christmas party was important not just to the project, but to the entire valley and its value was demonstrated "when Mrs. Lee Harrison mushed in 28 miles from Chickaloon to report the number of children at that end of the valley.... (There was no road to Chickaloon.) She was willing to walk in and report so that the dozen kids of the distant camp would be assured of a good time at the Christmas party."[4]

We invited Ross Sheely and his two boys over for Christmas day. Mrs. Sheely had health problems, and she and her 13 year-old daughter were in the States for medical treatment during the Christmas season. I figured that in addition to helping Sheelys through the holidays, having them over would help us overlook the fact that it was the first Christmas in 12 years that we weren't with our family.

While planning for Christmas continued, life went on. We celebrated Priscilla's birthday (December 10), did domestic chores, attended school, church, and community meetings, but again, our main job was still finishing the house. By Christmas the interior was walled and the kitchen and living room floors were laid. By December the days were very short. The sun did not rise up over the mountains where we could see it until after

10:00 in the morning. As one woman put it, "We had indirect lighting." The sun shown onto the mountains on the western side of the valley, and the pure white snow of the mountains reflected a fine light into the valley. The sun set before 3:00, but when the weather was clear, the rosy reflection hung on the snowy mountains for over an hour and it was the most beautiful sight one could imagine.

Even with the short daylight hours, we still enjoyed outdoor activities such as ice-skating (there was a low spot in one of the fields near the house that made a dandy skating pond), and skiing when there was enough snow. We had plenty of hill for skiing, but scant snow, (the valley received less than 55 percent of it's normal .9 inches of precipitation in December 1935[5]), and the wind usually blew away what little snow there was.

Wind was a constant presence in the valley. Sometimes it was just a gentle breeze, but fierce winds sweeping down off of the Matanuska and Knik glaciers could ravage the valley. Jack Alman wrote about it in the December 5 issue of his paper. "Did you notice that the wind blew? Well, it did. Tents were ripped from their frames. A two-story building of bare studs and rafters collapsed on the lot in back of Bert's [store]. Tar paper was torn from shacks, and by the way, if you've missed anything...such as garbage can lids, sheets of celotex, coal buckets or washing-machine crates...just come down and look in the front yard of the *Pioneer*. A lot of stuff piled up here."[6]

When darkness forced us indoors, and homework and other duties were completed, radio was one of the main forms of entertainment. We bought a radio in early December, and some perfectly wonderful music came over the radio from the Anchorage station. Up until 10:00 p.m. we could get programs from the states, but most of them signed off at midnight, which was 10:00 in Palmer.

Radio was about Alaska's only means of getting news to folks in a hurry. Many personal messages went through every night from Anchorage. Letters often would take months before

Matanuska Valley Christmas

they could reach their destination and then might not be called for. But nearly everyone had a radio to keep in touch with the outside world.

High point of every student's day—the bus ride to and from school.

Neil was still driving the school bus for that section of the valley, although he and I were quickly tiring of the job. Some women might have been called "golf widows." By the same token, I could have been be called a "bus widow." Neil spent so much time at his bus driving duties, it was disgusting. He had the worst roads in the entire school route system. Men who lived on the worst strip of the road on the bus run, a section of Bogard Road, said they would not have been responsible for that bus on those roads for $25.00 a day.

One Monday he was stuck in the water hole near our house. The bus had frozen brakes and he did not even get to school. Tuesday, on the way home at night, the bus slipped off Bogard Road, and in getting the bus out of the ditch it was quite badly damaged.

Wednesday, Neil was in town all day and most of the night working on the bus. Then, Saturday, he had to go to town in the morning to get another bus. Sunday he had to make the church route. On the way home, the fuel line bothered all the way and we spent about three hours on the road. Every so often Neil would have to get out and suck out the fuel line.

Neil went to school the next day in the same manner, stopping all along the way to suck out the fuel line. He said it mixed very nicely with his cod liver oil. He didn't go Bogard Road going over, but instead took a main road and skipped some of the children of the route. But the next night he came back Bogard Road and got stuck in a water and ice hole about 3 1/2 or 4 miles from our house. The kids walked home and were so tired. They just finished supper when Neil walked in. The bus was still in the hole and the fuel line had broken.

I had a committee meeting in Palmer that night for which transportation was provided, and Neil rode down with us to get help. He got home at 4:20 the next morning with his bus, after getting a tractor to haul it to town to get the fuel line fixed. Oh, it was a great life!

The great life still included hauling water, since well drillers had not yet arrived at our place. I was coming back one day with my second two pails. It was so cold (temperatures during mid-December were about -20 degrees F.) and the wind was cutting across the clearing, so I was hurrying to get behind the shelter of our trees.

In my hurry I caught my toe on a root and went flat. It knocked the breath out of me and wrenched my neck and banged my jaw, but what hurt worst was to see the two pails upturned and the water all on the ground! The next day my neck was just stiff enough and achy enough to keep reminding me of those two pails of water I had so nearly home.

Hauling water did have some small compensations, though. Those walks to the spring, just Neil and I, were special moments. There was a certain enchantment in the unfolding about of the Alaskan winter nights—a feeling that we would never be the same again after standing on the hill top and watching the moon ride regally along the horizon, casting its pure untroubled light on to the crystal snow of the mountains, sharply contrasting with the deep indigo of the star-studded sky. Soft blue shadows reflected along the jagged icy peaks, and crept down into the valleys, growing denser until they blended in to the sharp black and white of the spruce-spired lowlands.

In spite of all my water troubles, you could not have pried me out of the valley, though! (And I was the one that would not have lived in a house without complete waterworks in the States!) But I was just optimistic enough to think that someday we would maybe have a well.

A Corporation well-drilling rig.

At Greenes, our near neighbor, they first had a digger who went down until he struck hardpan. They then brought in a small drill, which it kept breaking, and after seven weeks, they finally got down 72 feet; but no water.

Then they pulled the drill and promised to bring out a big drill. One of the big drills had just burned down, though, and we were so handicapped for lack of drills! It was quite a while before the Greens finally got their well. But I suppose we had to get a taste of "pioneering" some way!

A Creek, a Hill, & a Forty

The same week the drill burned down, I went to town just to get the mail, but I took my pack sack along in case there were packages to take home, and there were. I rode as far as I could on the school bus (that was about 1 1/2 miles), then walked the rest of the way (about 4 1/2 miles). Usually when I went to town I stayed on the main highway as there was more chance for rides. But there was a short cut called the "Old Fishhook Trail" that wound back through the timber and made the way to town a little shorter.

I had never dared go through there in the summer as I was always afraid of meeting up with a bear. But since by December they must all have been hibernating, I decided to go that way. It was a lovely trail with only a few houses on each end, near the main highways. But I enjoyed it. It led through heavy timber, with the spruce so big and tall and strong, and the birch so graceful. There was quite a bit of large cottonwood there, also.

The little woods creatures seemed so unconcerned at my passing by. White rabbits bounded away, but stopped a short way off to look back inquiringly. Little brown woods mice (different than I've seen before, with short fat bodies and short flat tails) were very numerous.

I saw what I supposed were ptarmigan (though they did not look like pictures of ptarmigan in the dictionaries.) They were beautiful—all white, with a black corner cutting diagonally across each side of the tail. Another bird looked much like a big crow, only it had a big white spot on the top of each wing. Quite a flock of birds flew up into nearby trees as I passed. They looked very much like female cardinal grosbeak, only they didn't have so much of a crest or so red a head and bill. They also had a very sweet call.

When I reached the main highway again, just outside of town, I was glad I had taken the sheltered trail since there was a stiff cold breeze blowing there. (Though it had been perfectly calm when I left home.) Before I was through with all my errands, my pack board began to feel rather cumbersome. Six miles looked like a long hike with that on my back. It was just noon so I decided to stick around until after lunch in hopes I would catch a truck. I thought the delivery trucks would leave town about one o'clock, if there were any going my direction at all.

So a little after one o'clock I started out, and walked every step of the way! I met six pick-up trucks, and one of the buses, all on their way back to town! The pack-board possibly weighed from 25 to 40 pounds when I started out from town, and by the time I reached home it felt like it weighed more. (You know in Alaska, everything grows bigger, and faster than anywhere else in the world.) I reached home at 3:30 just ahead of Neil with his bus. Not five minutes later, a truck whined past going up the hill to the Greenes. About 15 minutes later, the doctor and nurse on their way out from town to see a patient who lives beyond us, stopped in at our place to get warm before going on. I think fate had the cards stacked against me.

Finally, Christmas week arrived. Saturday, December 21, was when the Alaska Steamship Company, the Red Cross, and Sears Roebuck gave their donations. Money was also given locally for treats for all. Busses ran all afternoon and evening getting the crowd to Palmer and home again. I went in the morning to help decorate and get the gifts sorted. It was all more or less confusing. But the hall (the new, nearly finished community hall) looked lovely with two trees, one on each side of the stage, and the stage background. It was all so beautifully decorated with gift donations of decorations. Estimates of the attendance at the program ranged from 700 to 1,000.[7]

The arrival of Santa Claus and his entertainers was an annual occurrence in coastal towns where the Alaska Steamship Company's boats docked, but in 1935 the Alaska Railroad also brought the entertainers north from Seward as far as Palmer. How fortunate we were! That year the Alaska Steamship Company's Santa Claus and 25 entertainers gave shows in Ketchikan, Wrangell, Juneau, Sitka, Palmer, Eklutna, Anchorage, Seward, and Cordova.[8]

After an all-day train ride from Seward, the entertainers arrived in Palmer late in the afternoon December 21. They found dinner waiting for them, after which they performed at the community center. (The community center was also the school gymnasium, which was still under construction. Construction had been sped up so the Christmas program could be held there.) The entertainers' part of the program consisted of Christmas carols, popular songs, instrumental numbers, and Santa Claus distributing fruit, candy, nuts and toys. Santa and entertainers then left for shows at Eklutna and Anchorage that same night before returning to Seward.

After the entertainers and Santa left, the Junior Red Cross of Seattle and San Francisco distributed gifts from Sears and Roebuck to all children. The gifts from Sears and Roebuck were in gratitude for all the business the colonists had done with the mail-order firm since moving to Alaska. Jack Allman estimated that including gifts from the steam ship company, Sears and Roebuck, and other sources, there were approximately 2,000 gifts distributed at the program.[9]

I had to return to town the morning after the program to fill gift boxes for children at Chickaloon and Premier. Those communities were up the railroad's branch line and there were no roads to them. I had to walk all the way down, and before I got there I began to feel as though I were slowly turning to wood. I was lucky enough to get a ride home. Neil didn't know that I had to go back to town on Sunday, and he invited Dr. Albrecht to bring a bunch out to sing Christmas carols on Sunday night.

Neil finished putting on plyboard in the living room and Mardie baked a cake, and we hung curtains (Janell and I) and Priscilla stood by on the cooking and dishes. Dr. Albrecht and his wife, Reverend Bingle, Mr. La Walter, Mr. Halvorsen and four young folks who work in the offices, came out. We gossiped and discussed shows and books and sang Christmas carols, and by the time we had coffee and cakes, I was all relaxed and in a very good humor again.

I baked all day Monday and Tuesday and cleaned and decorated the tree. Tuesday evening, Christmas Eve, the kids and Neil had to go to the program at Wasilla, but I stayed home to finish up. Each girl had on a "new" Christmas dress. Mardie's was a hand-me-down from LaMay Redmann at Blair, Priscilla's was a "made over" dress from Lucille Stumpf, and Janell's dress was the one I made for Mardie out of Mother's blue coat, and Priscilla had worn also.

But they looked so nice! I put a bit of rouge on their cheeks and lips to counteract the glare of the stage lights and I was almost breathless with pride when I saw them all ready. When Neil and the girls got home at 11:30 I was just finishing the supper dishes and still had the candy to make, the potatoes to peel, the chickens to clean and prepare. (There were no turkeys available in Palmer.) But we decided we would leave it until morning when Neil would help.

Christmas morning was a festive occasion. We didn't give many gifts to each other. Everyone considered being in Alaska gift enough. The girls exchanged paper dolls, and each received a watch from Neil and me. I gave Neil a box of nut-filled chocolates (my favorite), and he gave me a box of soft-center chocolates (his favorite). The many gifts we received from friends and family in the Lower 48 made up for the few we gave ourselves. Many colonists had a merrier Christmas than might have been, because of the generosity of people in the States. The top-notch present for me, though, that surpassed all others, was Arabella's gift of a dandy little heifer calf born December 23. Beatrice, Beulah, Beola and Beauty in turn were suggested for a name—but Ross Sheely's suggestion of Belva held the most favor. [The name Bozy won out.]

Sheely and company arrived at 1:00 p.m. for Christmas dinner. When they arrived we had not even changed our clothes. But they were so nice about making themselves at home. The Sheely boys helped turn the freezer and set the table, Ross mashed the potatoes, and Mr. Oldroyd (who was visiting the Sheelys and came along) stirred the candy. After lunch, the kids went skating. Ross helped with the dishes, and Mr. Oldroyd and Neil stirred candy. By the time that was all complete, it was time to get supper and milk the cow. I made the soup while Mardie and Jack (15) went to borrow crackers up at Greenes. The Greene's place was only across the 40, but Mardie and Jack had to take the Sheely car. They had more or less difficulty getting up the slippery hill, but both seemed to feel it was much more exciting than walking. Everyone else went over to see Arabella and her baby.

After supper (chicken and rice soup with orange ice for desert. As I told them, just water with chicken flavor for the first course, and water with orange flavor for the last one) we sang songs. Jack played the piano. They stayed until about 10 or after. Even though we were tired, we'd had such a good time we hated to have them go. After they'd gone Neil stacked the dishes and cleaned up a little, but I went to bed, with that "oh so comfortable feeling" of knowing that at last I could let down and rest.

We continued to receive gifts from friends and family in the Lower 48 as ships arrived in Seward. The stream of gifts, some from only nodding acquaintances, amazed us. But it let us know that we were not forgotten although we were far away. The Christmas program, and the gifts from people back home cheered all the colonists and boosted our morale during what could have been the loneliest holiday season of our lives. One colonist, Pat Hemmer, said, "This is the first real Christmas we and our children have had for five years, and we appreciate it."[10]

Nine
Mid-Winter Progress

Of the original 203 families in the project, we had dwindled to 163 by January 1, 1936. Eleven houses stood vacant waiting for new occupants. Sixty-two houses were completed both inside and out. The barn building program was finished for the winter, with 107 barns completed. Almost all the tracts had wells, and crews worked night and day drilling wells for the remaining tracts.

Community Center in 1936. School to the left (with gymnasium/community hall out of photo to the left. Staff dormitory and staff housing to the right.

Work progressed on buildings in the community center. The new school and community hall were nearly complete, and staff residences and a dormitory were being constructed. The community center sewage disposal system was finished, and the water mains were 90 percent complete. Workers were erecting the community water-storage tank.[1]

In addition to the Corporation facilities, private business was growing in Palmer. The community had a drug store, department store, several restaurants, a hotel, a taxi, one lawyer, and an insurance agent. The town's businesses saw a bright future for the community and formed the Palmer Chamber of Commerce in late January.[2]

A private trucking service running between Anchorage and Palmer started in mid-January. Since the highway bridge over the Knik River was not yet finished, the trucks evidently used the railroad bridges over the Knik and Matanuska Rivers, which resulted in citations for trespassing on railroad property.[3] Ernie Kling, who also operated a taxi in Palmer, ran the trucking service. Kling made weekly freight runs between Anchorage and Palmer charging $.30/ton/mile.

Several weeks before the trucking service started, however, the Reverend M. J. Jackson (Seventh-day Adventist minister), his wife, and two companions drove the first car from Anchorage to Palmer. They considered the $40 to transport the car by train too steep, so they made the 4 ½ hour trip following roads when they could, and dog sled trails when they had to. Ropes and blocks were used to traverse several glaciated sections of trail, and the Knik River was crossed on the ice below the unfinished highway bridge.[4]

The Maningens in our camp almost had the first baby of the new year, but their new one decided to come on December 31. Vivian had been ill right before the baby came, but tried to hide it. When Walter finally found out and called the hospital, she became upset. Dr. Albrecht came out with an extra man, but it took all three men to get her in to the ambulance, she was so hysterical. Albrecht said he thought it was just fear, since Vivian was so young. She wasn't even eighteen yet, and there she was with a two-year old and new-born. After the delivery Vivian calmed down completely, and she and the baby came home the first week in January.

Most of us colonists were clearing land, finishing our houses, or working for the Corporation, but a few "dissidents" still refused to work or improve their tracts. In a *Valley Pioneer* editorial January 9, Jack Allman wrote about the work situation in the valley. The ARRC originally opposed colonists taking jobs away from our farms, fearing that those of us who did so would neglect improving the tracts or abandon farming entirely. It also

Mid-Winter Progress

realized that paying jobs could not be provided for every colonist in the project. But the Corporation found it impossible to keep colonists out of the local job market. Most of us did not enjoy living constantly on credit, and we wanted cash for items not available from the commissary.

We were no different from other colonists in wanting items unavailable from the Corporation. We got so hungry for fresh vegetables, which couldn't be purchased through the commissary, that I sent to Anchorage for lettuce. The lettuce was $.20 (Lettuce was advertised in a Seattle paper at two heads for $.05.), and the express on it from Anchorage was $.60. That was a rather high priced salad, and we used it sparingly. If we wanted to buy mayonnaise it would cost $.80 for a quart bottle of Miracle Whip, which we used to pay $.35 for in the States. However, I made my own mayonnaise, and it wasn't so high priced, since I bought the necessary materials from the commissary and things there were very reasonable in price. It was only when we patronized the Alaska stores that we noticed the high prices.

Some colonists were employed by the Corporation, much to the dismay of others without jobs. Employed colonists were accused of playing up to Corporation officials and being administration favorites. Although Neil was employed by the Territory and not the Corporation, we were also accused of playing up to administration officials, and suffered socially because of our friendship with project officials. Neil's employment, and our friendship with Ross Sheely even became campaign topics during the community council elections.

Ross mentioned that anyone even friendly with project officials was suspect in some circles, and he felt the effects of this, too. Some colonists were hesitant to approach him for fear of what others would say, and he was hesitant to socialize because he didn't want to make trouble for anyone he befriended. As a consequence, he felt rather isolated at times.

Colonists with Corporation jobs received 25 percent of their earnings in cash, with the remaining 75 percent applied to their accounts with the Corporation. Neil's job was outside the Corporation, and although we didn't have to contribute any earnings, we voluntarily paid 75 percent towards our account. The ARRC also said colonists could earn money by cutting wood for mining-timbers or firewood. Colonists could keep all cash for selling wood to outsiders, and 50% of cash for selling wood to the Corporation.

In an editorial, Jack Allman said most colonists were working, either on their tracts, or for the Corporation. The colonists clearing land he referred to as, "real farmers who see more money to come out of the soil than they can earn with their hands." He also lauded those finishing their houses rather than paying carpenters to do the job, or working for the Corporation clearing land or finishing houses.

But Allman had harsh words for the few who "neither clear, build, or work, but shout to high Heaven that it is a crime they are being asked to sell the sweat of their brow for a mere 25 percent of a day's wages. They demand it all. Where, we ask, is the utopia where one can live on credit and sell his labor for cash?"

I certainly agreed with Allman. Anyone who followed the program set by the Corporation was allowed to work. They had to shut down on some as they would not do as the Corporation asked. They wanted all the benefits without doing their share of the work. Many colonists who followed the Corporation's program had good jobs and steady work, all paid on the Alaska relief scale—$1 per hour for skilled labor, $.75 an hour for semi-skilled labor, and $.60 for common labor if they were working for the Corporation. One of the colonists was marshal, one was commissioner, one was the shoe-repair man, one was a blacksmith, and so on.

I knew of several families having no intention of staying with the project permanently. They were just staying through the winter since they didn't have to pay rent. As soon as spring arrived they planned to leave. Any time the authorities got information about such families, they stopped all credit until they found out if the family was staying or not. If they were working and clearing land or showed signs of really carrying on when the matter was investigated, they got the credit returned. In a few cases they admitted they did not intend to stay and then were advised to leave immediately instead of later, and credit was not returned.

There was definite dissonance and dissent in the valley, some of it seemingly caused by a lack of information about administration policy and other matters. We often got information on the project, whether accurate or not, not from project officials, but from newspapers and radio broadcasts. Unfounded rumors circulated freely among the colonists, often being picked up by the press and publicized as fact. The rumor mill finally elicited a response from the *Valley Pioneer*. A January 16 editorial warned of the harmful effects of unfounded rumors. One of the rumors bringing this response was that the Corporation was requiring each colonist to clear 10 acres by May 1. (The story was picked up by the Associated Press, broadcast on the Anchorage radio station, and reported in the Fairbanks newspaper.[5])

Colonel Westbrook in Washington D.C. did notify the ARRC that clearing needed to be stepped up and that it was up to the ARRC board of directors and project manager to get it done,[6] but Allman reported that Ross Sheely denied he had demanded 10 acres cleared by everyone. Ross rather said he would like every colonist to clear 10 acres by spring, but it would be unfair to order a specific acreage to be cleared as some tracts were easier to work than others. "Because of the obvious unfairness of such a ruling," Allman said, "there is little doubt but what it has bred a feeling of dissatisfaction in the minds of many of the colonists who have heard or read about it."

Colonel Westbrook's telegram and the rumor evidently did have some effect on the colonists' clearing efforts. I didn't get away from home for so long that I noticed it myself, but some people from our camp said they saw men out cutting and clearing that they never saw working any time before. The colonists' information gap may have been due in part to a lack of communication between the ARRC and FERA officials in Washington, D.C.. Evidently, ARRC officials were as much in the dark on some subjects as were the colonists. (Ross Sheely wrote January 16 to Luther Hess, vice president of the ARRC board of directors, asking what Hess knew about the colony being taken over by the U.S. Resettlement Administration. Ross had read about the possible take-over in a newspaper.[7])

The colony council, formed to deal with colonists' problems and act as a funnel for communication, did little to calm dissent. In fact, one of the dissidents, Don Lund, beat me in the December council election. Although I enjoyed being on the council, I was just as happy that I lost. Holding office demanded so much time, especially when most of the time I had to walk to and from town.

But I was not very pleased with the man who did get the office. He was such a flighty, inconsistent (and what Arthur Stringer calls) "malcontent." I really believe he caused more discontent and restlessness than anyone else in the colony.

He had a gift of gab that made him sound as though he knew what he was talking about, but when you got to analyzing what he said there was no point to it. Most folks didn't analyze, they just took it as it sounded. He had been on the council, and every meeting he had some complaint. One meeting it was one thing, the next he would be standing just the opposite. He had officials disgusted all the time.

Of course, that was my opinion, and Lund did get enough votes to be elected, but Lund gave officials few further problems. At the first council meeting with the new members, he announced he was quitting the colony and returning to the States. The *Valley Pioneer* reported that by January 16, he had left the valley.[8] Guss Raschke replaced him on the council.

Colonists continued to steadily leave the project. Seven families left during January. A few families that we knew asked for return tickets in January. But we were not surprised. We had been expecting just those families to pull out—and they probably said the same sort of stuff the rest did. (You notice that most of those who had to return because of illness did not tell the same story as the others.)

By the end of January, an average of two families were leaving each week. Most of those leaving were from the same county in Minnesota (St. Louis County). It seemed they had not been satisfied any of the time since they had arrived.

There were 15 families sent from that county and most of them had the idea they were to be given good high-salaried jobs when they got here, and when they discovered what most of us already knew, that it was an agricultural project, with farming the main issue, they became dissatisfied and were discontented most of the time. Very little seemed right to them.

It seemed rather too bad. To those of us who were staying, it looks as though they were leaving before the real battle was on, and they were going to miss out on the greatest pleasure of all, that of seeing what they could develop from the fine start that we had been given. Some of them had their homes all finished inside and beautifully furnished. I should think they would have hated to leave, but most of them seemed glad enough to be going.

For some of the returning colonists, the lack of wells was the last straw. While most colonist tracts had wells by January, we and about 15 other families did not. Johnsons and we were the only ones in Camp 7 without wells. One of the drills was still at the Johnsons' tract adjacent to ours. The drillers gave up on their first well, and were having problems with the second attempt. Drilling crews were working 24 hours a day throughout the valley.

A major problem for the drillers, besides working in freezing weather, was the variable water level in the valley. Some wells came in at 20 or 30 feet, while others struck water at 90 feet, or were dry holes. Some wells dried up and had to be re-drilled. Several of the wells in the community center went dry, and new wells had to be drilled.[9]

By the middle of January our water situation was getting serious. We had to go farther for water as the spring creek had frozen so much that the water ran too slowly. In order to get water there Neil had to lie on his stomach and reach down through a 3-foot "well" of ice with a dipper, and wait between dippers-full for water to seep through. It was too slow business, so we were forced to go to the Wasilla Creek farther away.

We could get water at the Greenes, nearer, but they had so much stock and a not very satisfactory well. It pumped dry before they could get all they wanted for their stock, so we hated to take their water. The drillers were still at Johnsons having difficulties. They struck rock each time at 45 feet and had to change holes.

The drilling crew finally arrived at our place January 18. They brought their own lunch and coffee. I cooked their coffee and they came in to eat lunch, and when any of us women saw the sort of lunch those poor fellows had to eat, we got soft-hearted and gave them something different. (Probably that is why they had the women cook their coffee.)

The day crew ate at noon and the night crew ate at 9 p.m. I couldn't let the other women of the neighborhood show me up so I baked extra to give them at least as much as they got at Fanny's. (And you can believe too, the well drillers spread it on pretty thick about how wonderfully they were treated at Fanny's and the other neighbors!)

The well drillers were still at our tract a week later, struggling with unstable soil conditions. They had struck quick sand several times and the drill was between 45 and 50 feet down. One day the driller on the night crew made a wager that if we didn't have a well by that night he would bring out a box of candy. We didn't get the well and he brought two boxes of candy. Then the tool-dresser said that if we still did not have a well by the next Saturday he would buy the candy (providing he could win them on a punch-board like the driller did).

Mid-Winter Progress

The next day brought jubilant news. We had our well! I knew how a father feels who had waited hours on hours and at last had the doctor come out of the closed door to announce, "You have a son!" All day they monkeyed with the quicksand in the well—one minute thinking they had it cleared—then thinking that maybe they would have to dig deeper. But when Mr. Kelso, the foreman, came out that night to bring the night crew, he came to the door and in a very professional manner (smiling with satisfaction at having good news) said "You have a well. It is a very good well—inexhaustible." Boy! I don't know when I had such good news! Finally I could have a real honest to goodness bath!

While wells were a major concern to all the colonists lacking water, others in the project had different problems. An outbreak of smallpox in the valley and Anchorage raised concerns among Alaskans and those outside who read about the outbreak, but the few cases diagnosed were caught at an early stage and an epidemic was avoided. Only three cases were discovered in Palmer, and those were isolated to workmen from one bunkhouse, but the valley was quarantined just for good measure.[10]

The vacation resulting from the quarantine gave us more time to work on the house. Neil had all the ceilings and walls finished except for one corner of the dining room and parts of the kitchen. Floors were all laid except for the dining room, which was left unfinished waiting for cement blocks to build a chimney. Metal chimneys installed originally in all colonists houses proved to be hazardous and had to be replaced.

We finally moved our beds into the bedrooms and it certainly was nice. It was the first time since we had moved to Alaska that we had not slept all five in one room. It was probably largely imagination, but it seemed easier breathing then. I also had one of my built-in cupboards in, also a marvelous relief after being crowded for cupboard space so long. My relief at having the house nearing completion was short lived, though.

More work presented itself when we soon found ourselves building a shelter for the cows, which had been stabled at the Johnsons, our neighbors. Johnsons purchased another cow and wanted to get a team of horses, but their barn was only 16' by 20', too small for Johnsons' and our stock.

Harold Johnson built a shed for his sheep and pig, but even so, the barn would have been too crowded. So before Harold could get his cow, Neil had to build a barn for our stock.

We did not want to build our barn until after the frost was out of the ground in the spring as Neil wanted it in a side hill. So we decided to build our brooder house and use it for a barn temporarily. So Neil rushed with that during the daylight.

I hurried as fast as I could to get my work out of the way in the house so I could get out to help him, but even with Mardie and Priscilla on the job helping me (and Janell helping Neil) still I was not anywhere near through by noon when the girls had company over. Well, we were glad enough to see the company come, but it certainly shot the work help to pieces, and I did not get out to help at all.

When Harold went to get the team, they did not want him to take more than one horse. The horses made such a nice team (quite a lot of the horses shipped in weren't so good) and Harold didn't like to see the team "broken" so he induced Neil to take the other horse. So then we had a horse. Harold kept both horses until Neil needed ours, but he could not get his cow until Neil moved Arabella and Bozy out of the way. It was just a lucky break for us that we had the small-pox quarantine in effect.

Neil worked on the temporary barn during the days and puttered around the house in the evenings. He managed to get all the ceiling and walls finished, and was starting on trim work around the windows and doors. The interior doors still had to be hung, the linoleum in the dining room needed to be laid, and the entire interior needed painting.

The quarantine was lifted Friday, January 24, so the girls started school again the next Monday. Mardie was cramming for

Mid-Winter Progress

mid-year exams and lamenting that she did not get at it earlier in the vacation. She also washed and pressed her slacks to wear with a new "man-style" shirt to be like the rest of the girls in high school. They all wore breeches, overalls, or slacks, with boots, pacs or mukluks. So considering it all, I could say she was developing normally. One day was a day for sophistication, while the next was for paper dolls.

By the end of January our new "barn" was completed and the cows were home. The quarantine was over, everyone was back in school or working, and meetings were again allowed. Life was once again back to "normal."

Everything was normal except the weather. Neither we nor any of the other colonists had any idea what a normal Matanuska winter was like, but the winter of 1935-36 was slightly warmer and dryer than usual. (January's precipitation was 75 percent of its normal 0.9 inches.[11]) It was rather a joke. Here we were in Alaska all set for winter weather, lots of it, and we had spring weather while the folks back in Wisconsin had blizzards, tornadoes and sub-zero weather. On many days it was at least 40 degrees above. Practically all month we had such glorious weather, most of the time balmy. The day the eastern part of the U.S. was having blizzards we had a lovely soft fall of snow. Then the day the western part of the U.S. had its troubles, we had rain and melting weather that took most of our snow away again. It was with difficulty that I could persuade the girls that they needed to wear coats and hats when they went outside.

Ten

A Marvelous Winter

Social activities surged at the end of January with the lifting of the valley's small pox quarantine. The first weekend in February the band, and boys' and girls' basketball teams from Wasilla traveled to Anchorage. Neil went as substitute basketball coach, and Mardie earned her way by working around the house for me.

Neil and Mardie both had a glorious weekend. Mardie rode on the winged steed of happiness for two weeks because of that trip; the week before the trip, in anticipation, and the week after, in retrospection.

Winter view of Anchorage downtown.

They left Wasilla at 10:15 Saturday morning and came back in the late p.m. of Sunday. Railroad and hotel rates were lowered for the occasion. Mardie bummed around with one of the Wasilla girls Saturday afternoon, and bought Priscilla and Janell gifts. Then in the evening the basketball games were played. Wasilla lost both (40–7 and 37–5). After the games they all went to the broadcasting station where the Wasilla orchestra put on a program.

Neil and the Storms (friends in Anchorage) went to the dance hall after the program, and Mardie was quite thrilled to be invited along.

"Night life" was something so entirely new to her—she sort of felt at last she was "grown up" I guess. She even danced twice with Neil. Then they went back to Storms'. They apparently had a very thrilling time according to Mardie's enthusiasm. The highlight came in that they didn't get back to the hotel and to bed until 4:30 Sunday morning.

Mardie's roommate wanted to sleep until noon, but Mardie did not want to miss so much time. She figured she could sleep at home, so she was up early window shopping with Neil. Then at 10:30 the 12-year-old Storm boy was supposed to take her to get a chocolate soda (arranged by Mrs. Storm the night before), but when he came, he was accompanied by his 15-year-old brother, a friend of his, and two of the older Wasilla high school boys. She was quite proud of the fact that while all the other Wasilla girls had boyfriends for the dance the night before, she went to have a soda with five!

A major social event in Palmer was the dance on Saturday, February 8, to dedicate the completed community hall and crown Palmer's entry in the Miss Alaska contest.[1] The radio announcer at the dance said it was the largest and most modern community hall in Alaska. I don't know if that was really true or not, but I wouldn't have been surprised. It surely was a dandy big hall—steam heated, electrically lighted—with big balconies all around (and the place was all set to install our movie outfit, with sound projector).

The dance was a huge success. Everyone in the valley was invited and 800 people attended, including 215 people from Anchorage.[2] The orchestra for the evening also came from Anchorage. Practically all the buses, trucks, and taxis in the area were busy most of the night ferrying passengers to and from the dance. (Ads in the *Valley Pioneer* show at least three taxis operating in Palmer by February: Palmer Motor Service, Bob's Taxi, and Ole's Taxi[3])

The big attraction at the dance was the crowning of Miss Palmer. Every $.25 purchase at Palmer businesses, or payment of a bill, entitled a man to vote for his favorite contestant. The contest narrowed down to a neck and neck race between Virginia Berg and Bee McNally, both employees of the ARRC. Each ticket to the dance also entitled the men to four additional votes. Proceeds from ticket sales were to send Miss Palmer to the Miss Alaska contest at the Fairbanks Ice Carnival in March.

We heard a rumor that Bee McNally was ahead, so it was a surprise when Virginia Berg was announced the winner by about 600 votes. It seems that one of the town's merchants had kept a sizable number of votes back until the dance, wanting to liven up the evening. McNally's supporters demanded a recount, but Berg remained the winner.

Neil and I certainly enjoyed the dance, but I found that after months of staying at home, my feet were out of shape for dancing. There were several of us who vowed to wear our pumps and high heels more often before the next dance to sort of "keep in shape." We found that a night in high heels and narrow toes wasn't so pleasant following months of wearing shoe pacs—with too much toe space and no heels. Almost anytime during the latter part of the dance Saturday you could go into the balcony and see women sitting with their pumps kicked off, resting their toes.

The next social event was the President's Birthday Ball on February 22.[4] The dance was held on Washington's birthday rather than President Roosevelt's birthday (January 30) because of the smallpox quarantine. Although we had the ball rather belatedly, it was about the nicest social function that year. There was a very nice crowd and everyone was well-behaved. People in Alaska usually danced until 2 a.m., which was a bit too late for some of us who normally went to church the next morning, but let it be recorded to their credit, there was a record crowd out on Sunday, and the majority of them had been at the dance.

A Marvelous Winter

Our own Palmer orchestra [organized by Father Sulzman] played and everyone seemed to enjoy it more than the out-of-town music. There was quite an assortment of types of dress. There were some dressed in gorgeous evening dresses and some in semi-formals, some in afternoon dresses, in "all occasion" silks, in sweaters and skirts and so on. Yes, and some even in the ever-present breeches and boots. (I got so tired of seeing them everywhere, but still, I was guilty of it at times myself.)

The valley continued to have moderate weather in February (66 percent of normal precipitation[5]), which I considered fortunate. In early February I left milk outdoors every night for weeks and didn't have it freeze during the night. During the day it was about 20 degrees above and the sun shone with no wind. So it seemed "spring."

By mid-February the weather had turned slightly colder, but day-time temperatures were still above zero. The valley also received more snow. Not much, but enough to remind us it was winter and that we were in Alaska.

Old timers said that winter was not typical of all winters in the valley, as the roads in the valley were known to be blocked with snow. But with all the transportation of building material, feed, and such that was necessary that winter, it surely was fortunate we had the mild winter we had—also in view of the fact that many of the houses were still not complete. In some of the log houses, insulation was to come from dead air space between the logs and plyboard, and the plyboard was not up. It was a godsend that the winter wasn't severe.

Some women located in the lower parts of the valley near the river said they had not been able to get any clothes dry outdoors all winter due to the thick frost. But in our camp we dried all of our clothes outdoors. The wind at our camp not only snapped the clothes out, but also increased the temperature.

But you could never get anybody to agree that anyone else's part of the valley was better than their own! Folks in Camp 8 and 9 (near the river) could not be hired to live in Camp 7 where the wind blows so much. And we in camp 7 did not want to live where there was so much frost! And we did not want to live at the Butte, tucked in at the north foot of the mountains, where the sun shown so scantily. And the Butte folks thought their location was the "real pick" of the whole valley because the warm breezes from the inlet caused vegetation to mature more quickly than elsewhere.

Although the weather was moderate, and February's social activities were a welcome break from the mid-winter doldrums, families disillusioned with the project continued to leave Alaska. Seven families left the project in January, and another four left in February.[6]

I knew there would be quite a large number even yet to leave the colony and go back. I could not possibly, even by the wildest stretch of imagination, get their viewpoint. But with some it was quite acute. They did not have transportation or opportunity some way to get out and mix socially as they used to back in the States. They bought much trying to satisfy their feeling of discontent. The more they bought, the more they owed, and the more discontented they got over how large a debt they had. It was really rather pitiful.

Skating rink between the staff dormitory and the school. One of the first sports organized in the winter of 1935-36 was ice hockey.

A Marvelous Winter

With some it was just homesickness, and if they went back before May they had their transportation paid because they were still legal residents of their home states. After that they were Alaskans and had to pay their own way if they went. At that point I would not have been surprised to see four or five families leave our group in Camp 7.

With most of them it was what was called "cabin crazy" that was causing it. They never got out and attended church, or took part in any of the community affairs.

Of course, transportation was the big drawback. But they were not interested in going to church and that was the only thing for which regular transportation was provided. The community hall did not have a good floor until February, so there were too few public gatherings.

But of course some did not dance and lots of them had no one to leave with the children, so they were just tied at home. I don't imagine it helped any to see some of us on the run all the time. You saw the ones who were working in the church, the Legion, and Auxiliary finding ways to get out socially (if not in public gatherings, they made their own in their own neighborhoods), and working in the women's clubs. For the most part you met the same group in all activities.

My weekly schedule usually included at least two meetings. I was secretary of the American Legion Auxiliary; a member of the homemakers' club, our church board of directors, and the church choir. I also volunteered, or was volunteered, for other activities such as serving at American Legion functions, or planning for the community Christmas program and the President's Birthday Ball.

One of the construction engineers predicted that by May 1 only about 100 of the original families would still be in the project. He said that anytime a large group was called in to a new job, officials expected and planned for a 50% turnover in personnel—that only half the workers ever proved satisfactory. The engineer figured that our project was no different from any other.

A Creek, a Hill, & a Forty

With colonists departing, there were 17 tracts vacant by February.[7] Rather than wait for new colonists to be selected and brought in from the States in the spring, some of the vacancies were filled by Alaska residents. Nine new colonists drew for tracts on February 8.[7]

The day before the drawing, Matt Onkka and Betty Herman, grown children of colonists, were married. It was the first colonist marriage in the project. Only families were entitled to be members of the project, so their marriage enabled the new Mr. and Mrs. Onkka to qualify as prospective colonists.[8]

Project officials were apparently hesitant to allow the newlyweds into the project, fearing it would set a precedent for giving colonists' children tracts. But officials overcame their anxiety, and Matt was allowed to draw for a tract with eight other men at the Trading Post February 8. One additional colonist was added to the project by the end of February.[9] The new colonists were ecstatic at their selection. Some were homesteaders lured away from their own spartan holdings by the roads, stores, and other conveniences of the project. Others were workers employed by the Corporation. I talked to a truck driver who was to be allowed to draw for one of the tracts left vacant by returning colonists. He was so happy and so enthusiastic over having his chance.

Selection of the replacement colonists may have been as important to morale in the project as any other event in its early life. It showed that while criticisms of the project abounded, the colony had support. Some people still believed in its value and thought they and the project could succeed.

Jack Allman wrote in a *Valley Pioneer* editorial February 6, that the selection was a marvelous affirmation of the project. "All of them [new colonists] have had a splendid opportunity to study the set-up from all angles. They have heard the disgruntled rumblings of the dissatisfied and have heard the expressions of confidence by those who intend giving the deal a fair trial.... They have seen families return to the States with the lame excuse that the majority of the people were not the class they had been promised as future neighbors; and they have heard colonists say that they couldn't be driven from the country with a gun."

"Having heard both sides and seen all angles of the set-up as it now stands, these men are not only willing to assume the debt for the house and for work done on the tracts they will draw, but each and every one considers himself lucky to be offered the opportunity. What better answer to the howls of those who feel their return to the States has to be justified by adverse reports. What better argument to meet the long-distance criticism expounded by those who have never been in the valley and know nothing of its potentialities." put in letters received by Allman from people in the states.

Market Street in Palmer (now Dahlia Avenue).
Building on the right is the ARRC Trading Post.

Confidence in the project's future remained high among project officials, among most of us remaining colonists, and among Palmer's business community during February. The post office was raised from a third to second class office, and a new post office building was opened. A new railroad depot, with residential quarters for the depot manager, was finished, replacing the old box car depot. The University of Alaska opened a University Extension office in Palmer with an agricultural agent and a home demonstration agent. A restaurant, the James St. Clair Lake Resort, opened near Finger and Cottonwood Lakes. The resort featured dancing and dining, private parties, and public dances every Saturday night. Car service from Palmer was even provided. M. D. Snodgrass, a long-time settler, opened up a subdivision just north of the city center, and did a booming business. The Catholics bought two of his lots for their future church.[10]

Our church executive committee met to start plans for a permanent church building. The congregation was rapidly outgrowing the manse's meeting room, and we eventually had to move services into the school gymnasium.

February was a relatively festive month, with dances, marriages, and other activities. But a rash of fires, starting in January and continuing well into spring, kept us all a little sedate, and kept the volunteer fire department busy. Although a volunteer fire department had been busy for several months, the Colonial Volunteer Fire Department was officially organized February 18. Father Sulzman was the first fire chief, and he quickly drafted Jack Allman as the assistant fire chief.[11]

Walter Maningen from our camp, who had been snaking trees most of the winter with the Greene's horses, ended up in the hospital when one of the horses kicked him in the ribs. He had a one-month old baby at home, so the accident terribly upset his wife, but he was out of the hospital quickly enough, vowing to get back to snaking trees so he could get even with the horse. About a month later, to the horror of Walter, their three-year-old son walked right between the legs of the horse that had kicked him. The horse didn't even flinch. I guess it just had it in for Walter.

The Trading Post and other ARRC facilities started using "bingles" for transactions in February. Bingles were aluminum and brass coins, used instead of U.S. currency. The term evidently referred to poker chips and had nothing to do with Reverend Bingle. We, like every other family, were given our monthly allowance in bingles and could use them as we saw fit. Bingles were introduced to simplify Corporation bookkeeping, especially in matters between colonists.[12]

Colonists had little cash, so before the introduction of bingles, transactions such as property transfers between colonists had to be noted as adjustments to Corporation accounts, a paperwork chore. We could settle with each other using bingles, easing the Corporation's job. Bingles were to be used originally only in Corporation facilities, but it wasn't long until they were accepted by other commercial establishments in Palmer.

Palmer and the rest of the valley were certainly busy during February, but farms were still the center of most colonists' lives. Although we were rushed with school and social activities, most of our energy was still directed at fixing up the homestead.

By mid-February, Johnsons' house had been finished inside (carpenter labor at $1 per hour), even to the cupboards, and other cabinet work. All that was left was the painting. Johnsons were our nearest neighbors, and the inside of their house was built to our plans, so we could see somewhat of what ours would be like. Of course they made changes in the plans that gave theirs a little different look than ours. It looked so nice over there. They had all the rugs down, furniture in place, and shades and curtains up. Oh how I did envy their cupboards and drawer space when I had to look through boxes and trunks and barrels to find things. But I kept telling myself how I would appreciate all this more when it came time to pay interest and principle.

Neil devoted most of his time at home to finishing the house. The Corporation finished manufacturing the cement blocks for house chimneys by mid-February, and Neil had our new chimney installed by the end of the month. While Neil finished the house interior, I worked on clothing the family. Everybody in Alaska wore Montgomery Wards and Sears Roebuck clothes. I was determined to make ours for the sake of individuality as well as economy. Even if we had bought clothes at the Trading Post we would still have been dressed exactly like everybody we met. Only instead of being dressed in clothes of that year's make and that years catalog, apparently Sears and Wards figured it was a good opportunity to slough off all the old stuff they had on hand for years and never could sell.

You see, all government orders over $100 had to be put out to competitive bids, and if an order called for 100 dresses in assorted sizes and styles they got just that. What an assortment! They had to accept the lowest bid, and if you could see the stuff that came you could readily understand why it was the lowest bid! They really got in a nice line of men's clothing, but you had only to look at the line for women to understand just why we all wore our husbands' clothes!

The household and farm chores still demanded attention, of course. I didn't wash large items such as blankets, curtains, and rugs when we were still hauling water, but now that we had a well, I had to wash all the saved items. As a result, I had a regular "wholesale laundry" to do in addition to my regular chores.

I had become proficient at milking our cow Arabella since Neil drove bus on weekdays. Arabella proved to be an excellent milk producer, not always the case with other colonists' cows. Livestock was bought hastily at the beginning of the project, and the quality of the stock was not always the best.[13] Arabella proved to be an exception. After our temporary barn was finished, and Arabella and her calf back at our place, I lavished more attention on them, giving them warm water at noon and so on. Arabella noticably came up in her milk production then, providing at least 16 quarts of milk per day.

That was enough so that we could sell some of our milk in Wasilla. Mrs. Thuma (wife of the Wasilla principal) had a new baby and their regular milk man could not furnish enough milk for their needs, so they took two quarts a day from us at $.20 a quart. That sounds high priced for 1936 (milk was selling in Seattle for $.05 a quart.[14]), but when you consider that hay was $60 a ton, it wasn't so bad.

With the extra cash from Neil's job and selling milk, we were able to indulge the children somewhat. We let Mardie earn a trip to Anchorage doing chores around the farm, and the other girls demanded the same treatment. Priscilla spent her earnings on a permanent for her hair. Priscilla received her permanent, and I tell you, she was a happy girl. We had a beauty shop in Palmer—$7.50 for permanents. I imagine Priscilla would never spend another $7.50 that brought her the unqualified joy that did. All that pent up noise that she harbored for so long was finally set free. She plagued Mardie and aggravated Janell, and talked too loud and laughed too hard and ran too fast—until it was a bit overwhelming. But she sat for so many years with her nose in a book—too quiet, and too pensive—so it wasn't surprising that it finally burst out.

Although February brought an increase in community social activity, home entertainment was still paramount, and radio was a main focus in most homes. In the valley, as elsewhere, the radio reception varied with individual machines. We never took the time to really install our radio with proper grounding and antenna, so we got only Anchorage and the West Coast stations. But those who made an effort to really get results received almost any place they chose. At 11:30 every night it was very easy to get Paris, where world news was given not only in French but in English as well. According to the *Valley Pioneer*, the station was Radio Colonial, a service of the French government. Schenectady, New York, seemed very easy for most people to get. I even had it one night.

Schenectady was quite a ways from Palmer, and being able to receive news from there and other points in the U.S. and the rest of the world made us feel that we were not so isolated, especially during the small pox quarantine in January, when most of us could not even leave our farms. Events in February showed us that the isolation was only temporary, and that there was life in the project.

Eleven

Springtime in Alaska

March and April were fairly calm for all of us in the valley. The Fairbanks Ice Carnival March 5-9 was one of our main topics of discussion. Our primary interest in the carnival was the Miss Alaska contest. The competition drew contestants from all over Alaska, and even Dawson City, Canada, sent its queen. (She was there only as a representative and did not compete.) The Fairbanks newspaper wrote that Juneau's determined contestant mushed from Skagway to Fairbanks for the competition.[1]

A number of Palmer residents took the special excursion train to Fairbanks to take in the ice carnival and cheer on our contestant in the Miss Alaska contest. Those who went were not disappointed when Miss Palmer, Virginia Berg, was chosen Miss Alaska on the first day of the carnival.[2] A rousing start for the newest city in Alaska!

The new highway from Palmer to Anchorage neared completion in March. Although the bridge over the Knik River was not totally finished, it was able to handle traffic. The *Valley Pioneer* wrote that George Connors and Arvid Johnson, both of Palmer, drove the first automobile between Anchorage and Palmer using the new bridge.

While they may have driven the first car across the bridge, the first vehicular traffic was reported to be a dog team driven by Mrs. Lee Rees of Anchorage. She was following the new highway to Palmer and planned to cross the Knik River near the unfinished bridge, but she found the ice too thin to cross and the new bridge only about 3/4 decked. Mrs. Rees drove the dogs as far as she could across the bridge until she faced a 15-foot drop at the end of the planking. Fortunately, workmen were there and they assisted Mrs. Rees in lowering her dogs and sled to the ground.[3]

National interest in the project had ebbed somewhat over the winter but it was still sufficiently interesting to warrant an

Springtime in Alaska

NBC radio broadcast from Palmer. A radio crew set up a temporary broadcasting studio at the school building in mid-March, and project officials quickly prepared a program. I was invited to participate, and worked on the program with Jack Allman, the show's emcee.

The program was recorded March 13, but because of problems with the equipment's batteries, it was not broadcast until the 17th. I spent the entire day of the recording in town waiting to perform, but never got the chance. The program had to be cut and revised and cut again. NBC finally wired to the radio crew and told them just what was to be said and done, and they never got to my part.

The half-hour program featured music by the Palmer band (recently named the Palmer Pioneers), and speeches by project officials and colonists. Speakers included Ross Sheely, Dr. Albrecht, Reverend Bingle, and Jack Allman.[4]

Don Irwin, former project manager and now assistant manager, did not take part in the program, perhaps because of his recent decision to resign from the project. It was not supposed to be public knowledge yet, but we knew in mid-March that Don's time with the Corporation ended May 1, when he would return to his former position as manager of the Experiment Farm in the valley. His successor at the farm had not been successful and was not as well-liked, so Don had been asked to go back. His resignation was finally made public in the *Valley Pioneer* April 2 issue.

I took a short vacation from homemaking in mid-March and went to Anchorage with a friend. The trip was part pleasure and part business. I had been having recurring headaches and decided to have my eyes checked by an ophthalmologist, the nearest one being in Anchorage. The doctor told me that I had an astigmatism and needed glasses for close work such as sewing. While in Anchorage, I took in the sights and even splurged on myself—a rare event. For the sake of my family, I actually took time to get a permanent. It cost me $7.95 and was very lovely and soft, and gratefully welcomed by the family. It constituted my entire spring buying. My existing wardrobe had to suffice, and I kidded myself into thinking it was enough as long as my hair looked decent.

Anchorage at that time had a more or less a floating population, but the count on the ones who stayed around long enough to get in on the census was around 2,000.

The city did not look to be that large, but I think that was because of the little "doll" houses that most people lived in, which did not take much room. The houses in Anchorage were, I can think of no other word with which to describe them, they were "cute." They were so tiny. All of them were even smaller than the ordinary bungalow styles of houses so common in the States.

Fourth Avenue in Anchorage

The main street, Fourth Avenue, looked much like any main street in a little country town of about 2,000, but they did a tremendous business there, as they served so large a territory. They had many hotels and all seemed busy, caring for the floating population. Most of the stores and shops were as up to date as those Outside.

When we came up here so many folks thought we would never be able to have a good doctor, or dentist, or druggist, or what not, but Anchorage, I would say, was just as well supplied with all of such as Eau Claire Wisconsin was.

The school in Palmer was nearing completion by the end of March. Classes for the students that had been studying at home began in the building March 23.[5] Having classes in the unfinished building was certainly a relief for the school teachers who had traveled from house to house all winter.

The *Valley Pioneer* reported that additional water wells had to be drilled at the community center due to the increased need for fire protection, and to provide a water supply for the school.[6]

Mardie was also able to get away from the family in March, spending one week with friends closer to Palmer. She chummed around with her friends in town all week, making her sisters envious.

Priscilla and Janell [still attending classes in Wasilla] did not have the opportunity to be with the children of their age at Palmer except for the Sunday School hour. Priscilla said, "All the kids at Wasilla can do is look up nasty words in the dictionary."

They had quite a few nice friends through Sunday School, but no way of mixing with them socially as nothing much was done for kids their age.

The only other children in our neighborhood ranged in age from three months to five years. Priscilla and Janell played by themselves most of the time. Priscilla was a voracious reader and Janell was still young enough to be entertained by her dolls. Fortunately the winter was mild enough for them to play outside a great deal.

It was not until the end of March that the valley received any significant snow fall. While the rest of winter had below average precipitation, March's precipitation was above average, .512" compared with its normal .375".[7] The girls certainly enjoyed our snow when it finally came. We had enough to actually look like winter. It was quite soft and made good skiing. They even made a ski jump after supper one night. It was light enough to play outside until nearly eight o'clock. The sun rose about 6 a.m. and set about 6 p.m., but it stayed light longer, and it also got light about 4:30 or 5 in the morning.

I was able to channel some of the girls' energy into helping finish the house, but progress on the house's interior seemed agonizingly slow to me. However, it was nearing completion. Neil finally completed my kitchen cupboard shelves, and I was able to move my kitchen out of boxes and into the shelves.

By the end of April the house was tantalizingly close to completion. I really was beginning to have hopes. Neil finished the dining room floor and put in all the window frames and some of the door frames, and we had curtains in all the rooms but one. I had not finished my living room curtains, so we were still using the bedroom curtains there. I went through a process similar to house cleaning. I sort of took things out of one box and put them in another, moved the bunks around, washed windows, took off the storm windows, and other chores. Then I put up curtains; so

The Millers' first cow, Arabella

Springtime in Alaska

I could say my housecleaning was done for spring, though we continued building, putting in doors, putting up curtains, and moving boxes.

Taking care of our cow Arabella still took a considerable amount of time, and not a little worry. Arabella bloated at the end of April, and the man in charge of stock for the project had to be called. He forced Arabella to drink some kerosene and medicine and she got over the bloat fairly quickly. Arabella's milk tasted like kerosene for a week, but she was fully recovered.

We also waited anxiously for additional livestock to arrive. The Corporation received 72 cows in December, but they weren't acclimatized and many of them died. Additional livestock was ordered in March, including 7,000 chickens, 277 geese, 152 ducks, 225 turkeys, 550 sheep, 83 hogs, 144 dairy cows, and 114 horses.[8]

The livestock had not arrived by the end of April, and I was beginning to wonder if the animals would arrive at all. Rumors circulated that the order had been cancelled.

I thought chicken farming would be a profitable venture in the valley and was ready to order chickens on my own if the Corporation order did not arrive. By raising poultry we could always have a market, a good one, whether the project folded up or not. All the eggs we had since moving to Alaska had been shipped in, and they were not very tasty by the time we got them. And it was practically impossible to buy chicken to eat, unless storage ones were shipped in.

I felt daring enough in April to take a chance on what the *Valley Pioneer* referred to as a "darn good gamble,...the squarest there is."[9] We bought tickets on the Nenana Ice pool. One was not a real Alaskan if he didn't do that. Every year tickets were sold at $1 each, to guess when the ice would go out of the Tanana River at Nenana. The one guessing nearest correct won the jackpot. In 1935 the jackpot was won by a newcomer who had not been in Alaska much over a year.

A Creek, a Hill, & a Forty

We were still contributing 75 percent of the earnings from Neil's job to our account with the Corporation. I went in to Mr. Cronin's office at the beginning of April with $225 I wanted to apply on our indebtedness.

Cronin (fiscal agent) was in anything but a sweet humor. When he finally got around to ask me if I wanted something, he went on quite at length to tell me that I could not have it if I did because they had no money. Cronin was in charge of purchases for the project. Any request from colonists for items not available from the commissary went through him. Many colonists also regularly asked for cash advances from Cronin for items not available from the commissary that could be purchased either elsewhere in Palmer or in Anchorage. He always amused me, even when he was cross, and I told him he must not be in a very good humor, and he admitted he was not.

I showed him the check (I had two of them) and he said very gruffly, "I cannot cash them. I told you there was not any money!" You should have seen how "flattered" his face was when I finally made him understand that I was giving him something. He said he thought he could smile the rest of the day. I could see why some got so exasperated with him, but I couldn't help but get a bounce out of him.

I wasn't so pleased later in April when I went in to order living room linoleum and window shades. I still had between $140 and $200 left on our furniture allowance, but the Corporation was out of funds and I was told I couldn't get the linoleum and shades unless the ARRC received another appropriation from Washington.

We were not particularly concerned, and we did find out later that our linoluem and shades had been ordered, but you could see why so many got disgusted. Here we made an effort to save and not get things before it was necessary, and we also paid in 75 percent of all earnings, and when we wanted things we could not get them!

Others had all their carpenter work done, had a painter do the painting, spent way over their $285 furniture allowance, lived beyond their budget most of the time, took everything that came whether they needed it or not, paid nothing in, did no more clearing if as much as we had, and were still kicking about the Corporation.

You could see why there was so much beefing. That sort of thing happened all the time. It rather made us wish we had not been so willing to pay the 75 percent of our earnings.

We may not have been too concerned about dwindling funds, but project officials were. Ross Sheely travelled to Washington in April to request an additional $1.5 million from FERA. The funds on hand were low and would have lasted only until May 1. It would certainly have been a peculiar situation to face if there were no more funds, but Ross did come back with an additional appropriation.

The heavy snow during late March was the last gasp of winter. April brought warm weather and breakup. But early April, while the days were long and the sun warm, was not a particularly interesting time of the year. It was just that in-between time before spring actually arrived. The roads were muddy and rutted, and the melting snow left soiled earth and dirty water holes. The only hope, born on the kind spring winds and the sun's warm caress, was that the earth would soon be delicately green with leaves and grasses, and dotted with the tiny spring flowers.

Much of April was spent preparing for spring planting. I started cauliflower, head lettuce, cabbage, celery, and tomatoes in tin cans in the house and moved the cans from window sill to window sill to follow the sun. I used any excuse to work outside: feeding the cows, cleaning windows, working on retaining walls.

Breakup meant that roads in the valley quickly deteriorated. The roads out our way from Palmer (on the higher land) were not bad where gravelled, but the roads near the river were about impassable. One Sunday, our bus was the only one to get in with the church crowd. We got stuck on the way home, though. Where we had to turn around in one place the back wheels got off the gravel and we had quite a time getting back.

That same Saturday the whole family went to town. Neil had the old Ford fixed (as well as could be expected considering it's extreme age). Driving to town was clear sailing, it was all downhill, but coming back, when we started climbing, it was a different story. One thing about it, Neil never needed to worry I would ever take the thing and go someplace alone with it.

Spring usually brought an influx of job seekers to Alaska. I even received requests from friends in Wisconsin for information on job possibilities. I responded that the Corporation did not have much to offer in the valley as funds were so low. There was some construction work in Anchorage and some work at the mines. Alaskans seemed to rather resent the influx of folks looking for work as they felt there was no more than enough to supply Alaskans. However, nearly every boat seemed to bring in more folks and many of them did get work.

The influx of new people didn't affect Palmer too much. The valley was relatively quiet. Several dances and two marriages did liven up the place in March and April, and Palmer also received its share of disasters.

In March, Robert Canning, an ARRC worker, was killed at the sawmill at camp ten when he fell onto the saw blade, severing both his legs... That month, a staff tent in Palmer was destroyed by fire, a small fire in the school building caused minor damage, and in April a fire also destroyed colonist Clinton Johnson's house.[10]

The fire at Clinton Johnson's house emphasized the need for an improved telephone system in the valley... Each camp had only one or two phones at centrally located houses. By the time the fire at Johnsons' was discovered and someone reached a phone, it was too late for Palmer's volunteer fire department to save the house.

Springtime in Alaska

According to a *Valley Pioneer* article back in December 1935, over 60 miles of temporary telephone lines had been strung in the valley— first to colony camps and later extending to settlers' homes. Those lines were still in use by the spring of 1936.

The system, which was one large party line with a maximum capacity of 16 phones, worked fine when there was only one phone per camp and everyone was still living cheek-by-jowl in tents. However, as people moved from tent camps to houses, and as settlers were connected, telephones had been added to the system, overloading the circuit. Also, the increased distance between houses, could, as with the Johnsons, cause delays in calling emergency services. [11]

Old warehouse inferno

The most disastrous fire in Palmer's short history broke out at 6:10 p.m. April 20 in the old project warehouse. The fire spread rapidly and Corporation employees were drafted to help the fire department battle the blaze. The fire threatened adjacent buildings, and it jumped the road, igniting piles of lumber. The ammunition for the National Rifle Association was stored in the warehouse and ignited during the fire, shooting in all directions, but no one was hurt. Another injury was narrowly avoided when an overheated fire extinguisher inside the warehouse exploded and shot clear across the road and railroad tracks, landing on the depot platform and just missing a child. Residents of nearby tents were hurriedly evacuated.

Firefighters were able to prevent the fire from spreading to adjacent buildings, but the warehouse was completely destroyed. The worst loss was equipment for the project's dairy processing plant and equipment for the school. The U.S. Signal Corps station located in a corner of the warehouse was also destroyed.[12] Fortunately, the old warehouse was almost empty. Aside from the supplies and equipment stored inside, the loss of the old warehouse was relatively minor since the Corporation planned to build a new one.

It just happened that I had occasion a few hours before the fire to go all through the building with some of the staff. We were looking for cups that were supposed to be stored there until the home economics room was ready at school. We didn't find them, but we made a pretty thorough canvas of the place. In comparison to the way it used to look earlier in the year (before the new warehouse was built and before the commissary was used for storage) it was quite empty. The apparent cause of the fire was faulty wiring. Faulty wiring also caused small fires in the school and hospital. Further electrical work was halted until all wiring in Corporation buildings was checked and repaired.

The departure of colonists slacked off during March with only one family leaving, but six families left during April.[13] Three families from our camp: the Fitzpatricks, Maningens, and Fredericks, pulled stakes in April to go back to the States. Fitzpatricks had planned to go all winter and never took an interest in the project.

The Maningens nearly split up the previous summer. The McCormicks, who returned to the States last year, almost had Vivian talked into going back with them. Then during the winter, the Maningens had a run of tough luck and high hospital bills.

Springtime in Alaska

Vivian had her baby at the end of December, and right after that Walter was kicked by a horse and ended up in the hospital. He hadn't been released long until the baby lost weight and wasn't doing well. Then Vivian had to have an emergency appendectomy. Along with all that their heifer calf sickened and had to be killed.

We really expected the Greenes and Kleinpiers to leave too, though they had not indicated they would. Clarence Greene claimed he was staying, but he was very thick with all those who had returned already. Ken and Grace Kleinpier were very intimate with the Fredericks, and had moved to camp 7 from camp 4 to be neighbors with them. With Fredericks leaving, Grace Kleinpier told me that her husband was discouraged too.

I was horribly disheartened by the departure of Fredericks, Maningens, and Fitzpatricks. Camp 7 was one of the more distant camps from Palmer and there were only eight farms there, so having three neighbors quit the project with the possibility of more leaving made me feel somewhat isolated. I was not about to give up, myself, so I was eager to have the tracts occupied again, but no new colonists were added until summer.

By the end of April, 14 colonist houses stood vacant. The manager's, assistant manager's, and school superintendent's houses were completed and occupied. Workers were finishing the staff dormitory and some ARRC employees already bunked on the dorm's second floor. Work was progressing on the other staff houses and the school. Clearing had started on the creamery site. Officials were also planning to set up a local radio station. A gas station opened at the ARRC garage, a new restaurant was almost completed, an addition to the hotel was under construction, and Koslosky's Department Store was doubling the store's size.

The Palmer Athletic Association, a Valley Farmer's Club, Camp Fire youth group, and Homemakers Club were started in March and April.[14] Palmer was growing rapidly, preparations for spring planting were beginning to buoy our flagging spirits, and May would mark the end of the project's first year in Alaska.

Twelve

Birthday Celebration

May began as had many other months in the project. More colonists left, and those remaining worked on their farms. The Greenes decided about the beginning of May to return to the States, as I thought they would. That left only three of the original Camp 7 group, plus the Kleinpiers who had moved on to one of the vacated tracts. Everyone expected Klienpiers to go too, but I didn't think they would, at least not that spring. Alida Green would have stayed, but Clarence was lonesome for the "bright lights and pavements." I hope they were not sorry, but I'm afraid they probably were. We hated to have them go. They were awfully good neighbors.

One of the few plusses in the departure of our neighbors was the freeing of cleared land for our use. We, Johnsons, and Raschkes leased some of the land on the vacant tracts for our gardens and grain fields.

We also bought one of Green's pigs to raise for next winter's meat. The pig was so cute I knew the girls would become sentimentally attached to it before fall. Mardie wanted to raise the pig for a 4-H Club project. Janell decided to raise Arabella's new calf for her project, and Priscilla, not to be left out, hoped that our sheep would arrive soon so she could have an animal for 4-H.

The snow was finally gone and the ground thawed, so we started preparing fields for pasture and planting. Colonists had about 400 acres of cleared land by summer, 1936. Little land was cleared in 1935 because of the push to get houses and barns completed.

The Corporation encouraged us all to begin clearing early in 1936, and by summer about 225 additional acres were cleared. Including the 175 acres ready for planting when we arrived, the average acreage cleared on the 155 occupied tracts was

Birthday Celebration

Corporation Caterpillar tractor clearing stumps—1930s

2.5 acres, far short of the anticipated 8.5 cleared acres. Much of the land cleared in 1935-36 was not ready for cultivation by summer, though. Only 296 of the 400 acres of cleared project land was planted in 1936. An additional 225 acres of rented land was cultivated.[1]

We had slashed and windrowed only a few acres by May, 1936. Stumps, downed trees, and other rubbish from slashing and clearing was collected in long rows, called windrows. The windrows were then burned. In early May, Neil wanted to get rid of the grass to uncover all the downed logs, preparing to clear and fix pasture, so one calm morning he started burning grass.

I had washed out a few things and was only partly through scrubbing the living room floor when Neil called for help with the fires. The wind had come up and the fire needed more than one attendant to keep it under control. So we worked at that until nearly 2:00.

We had to go to town and did not dare leave the burning logs and stumps, so I got real daring and took the old jalopy, the family, Fanny and Jackie Johnson, and went to do the Saturday shopping, while Neil stayed to watch fires. I felt pretty smart to actually get there and back without any mishap, considering the age of the machine and that it had been four years since I drove a Model T.

One new development in the project that I was proud of was the establishment of a local radio station. The station, operated by the Army Signal Corps, began broadcasts in mid-May. It broadcast for an hour at noon on Tuesdays and Fridays and did not have entertainment or news, just information about the project. The idea was to provide the ARRC management with a means of communicating with the colonists, keeping them better informed about the project, eliminating rumors, and improving morale. Colonists were also instructed to tune in the project station after hearing the fire alarm so volunteer firemen could receive directions to the fire.[2]

It certainly was an excellent cure for many of the colony's ailments. It seemed to me that so much of the discontent and discouragement in the project had come from the fact that so many really knew so little definitely. For instance, during the winter we were sent questionnaires asking how many chickens, sheep, cows, hogs, and other animals we wanted for spring delivery. We filled them out and heard they had been ordered. It was not long until we "heard" the animals were not to be ordered as there was no money. Then we "heard" only horses were ordered. The next time it was mentioned we were given to understand that no horses were ordered, but the chickens were to be here the first of June.

Each informant was sure of what he said as someone in charge had supposedly given that information. So you can see how valuable a radio station was if it gave us the news, as we would know how to look ahead and plan for what was coming. We could also be aware of each change of policy in the management. [The radio station operated at least until 1939, but sometime between then and 1946 it stopped broadcasting. When the Palmer hospital was destroyed by a fire in 1946, the vacant radio station building became its temporary home.[3]]

Another more significant development was reflected in a telegram received from the Wisconsin Emergency Relief Admin-

Birthday Celebration

istration. The telegram stated that from May 14, 1936, one year from the colonists' departure from the state, they were no longer considered residents of Wisconsin, and would not be eligible for relief payments should they return. The Federal government also ended its policy of paying the return transportation for families departing the project. This was good or bad news depending on the colonist's viewpoint. The *Valley Pioneer* reported that some colonists greeted the news approvingly, since now if they quit the project they would not have to leave Alaska.[4]

View of Palmer from the west—Lazy Mountain in background (c 1936).

We were no longer viewed as legal residents of our home states in the Lower 48. We were quickly approaching the end of our first year in Alaska, and would soon become Alaska residents. The colony planned a big celebration to commemorate the project's first birthday. The American Legion and Auxiliary were placed in charge of the celebration, so naturally, I was in the middle of the planning.

The celebration was to be an all day affair (between chores). The project staff was also planning a dance in the evening. All of these big affairs seemed to always come on Saturdays, making Sunday a pretty hard day. I really rather dreaded Sundays, we were so tired, and had to get up and hurry so fast to get that bus off on time, then fight sleep through church, get home to a house without even a fire and have to fix dinner, until it left little time to relax. Still, all of these things seemed important to the growth of the community.

For the colony as a whole, we had taken a pretty big stride. The aim of the colony was to establish an agricultural community with cooperative marketing facilities. The cooperative marketing had not quite materialized yet, because we were not really into production. However, no-one could deny that the community was established.

Saturday, May 16, was the day set aside to celebrate the first birthday of the colony. It was called "Colony Day." If it did the rest of the colony as much good as it did me it was a truly important event. For several weeks before hand, I had felt sort of unwound. No particular reason for it that I could see, just a feeling of being easily annoyed and discouraged. Probably it was just a combination of having friends from this camp talking discontentedly before going back, together with the prolonged arrival of green in the trees, and too many events crowding the weeks prior to the closing of school. I seemed to be getting critical of everything and everybody.

I did not even want to go to the "Colony Day" picnic, but Neil was general chairman of the day's activities, which was reason enough that I should go. Then, some of the Auxiliary women had been asked to help pour coffee and serve ham, so I felt I had to go.

In spite of seeding season being on, with so many men feeling they had to stay home to work in the fields, there was a big crowd. I was surprised at how few of them I knew and how much the children had grown. When you gathered all the settlers, colonists, staff, and businessmen together it was a big crowd.

It was really rather amusing that when I first reached town I found three of my special cronies and the four of us com-

Birthday Celebration

pared notes. We found out we had all felt the same crabby way about coming, we were there under protest, and did not expect to have an especially good time. When the day was over we had all had such a good time and were in such good humor we decided to all come back to the dance in the evening. Some way, we all seemed to have taken a new lease on life in general, and everything seemed bright again. You know, even the trees were greener!

A program was held at 11 a.m. at the community hall, featuring speeches by Don Irwin, Ross Sheely, Governor Troy, Colonel Ohlsen, and also Dr. Ernest Gruening, Commissioner of the U.S. Territorial and Trust Possessions, and a member of the ARRC board of directors. (He became Alaska's territorial governor, and later one of Alaska's first U.S. Senators.)

Following the program there was a picnic in the large field in front of the school. The field was to be the community park, but then it was bare of grass or trees and still showing signs of construction activity. The ARRC furnished ham and coffee, and each family brought food to supplement the ARRC's donations. The day was sunny but chill, with a stiff breeze blowing off the Knik Glacier. Dust from the Matanuska River and from the piles of dirt around the buildings swirled through the air, but most of us were able to ignore the cold and dust and enjoyed ourselves.

After dinner there were races and contests for everybody: egg races, foot races, horse races (farm horses!), climbing the greased pole, catching the greased pig, a sack race, rolling pin throwing contest, and more. I won the husband calling contest. Very likely the reason for that was that the other husbands came when called the first time and I had more practice.

I would have liked to have had a movie of one little boy's activities. He was a kid about ten years old named Eddie, one of the kind with so much pep he didn't take time to do much growing. He was in every race and contest for which he was eligible and usually won 1st, 2nd, or 3rd in each one. He had on some old baseball clothes that were too big and he climbed the greased pole and won first. He chased the greased pig and was in the mass of kids that piled on top of the pig in the dirt. The pig had never been in a greased pig contest. It stayed in one spot and was mobbed by the kids.

Eddie then entered the pie eating contest. They had their hands tied behind them, and the one who finished his whole pie (blueberry) first won. You should have seen Eddie! He was just covered with grease and dirt, and blueberry pie in his hair, his ears, his eyes and face, down his neck and on his clothes. Oh, he was a mess! But he continued in everything that was going on.

The first person I saw when I went into the balcony at the dance that evening was Eddie—scrubbed! Scrubbed and washed until he must have fairly burned, and all dressed up, hair slicked back, a snappily pressed new brown suit, clean shoes and everything, and still wearing his grin and being a part of the whole affair. About 11:00 I ran across Eddie again, fast asleep in a chair in the corner.

The dance in the evening was free, with Palmer's Pioneer Orchestra providing entertainment. Everyone was presented with paper party hats, and punch and cookies were served. We left the dance about midnight, but our bus had a flat tire about a mile outside Palmer and we didn't get home until 2 a.m.

June 1 was to be Don Irwin's last day with the Corporation, and the high point of the Colony Day dance was the intermission when Don was pulled up on stage to receive some going away gifts. The project staff presented him with a pair of binoculars. All of the colonists chipped in and gave him a 12-gauge Winchester repeating shotgun. I wasn't sure what kind of gun it was but all the men seemed to be quite green-eyed with envy. So it must have been something unusually fine.

We, like many other colonists, were saddened when Don left the project, even if he was staying in the valley. Don sincerely regretted leaving, although his association with the project had not always been happy. The *Valley Pioneer* reported Don saying

Birthday Celebration

that his time with the project had been "one of the fullest of his life—fullest because of the many real friendships he made." The article finished by saying, "Good luck, good health and success express the sentiments of the whole valley who knew him during the most trying time the valley has or ever will experience."[5]

The board of directors for the ARRC met May 16 while the afternoon's festivities took place. It seemed quite appropriate to have it meet on Colony Day at the close of the year. Then we started afresh, with all things straightened out and decisions made, ready for the second year's activities. The sale of Corporation land to three religious groups was approved by the board at the meeting.[6]

The Seventh Day Adventists had already built a permanent church in April on land acquired from a settler.[7] Catholics and Lutherans were ready to build elsewhere, but the ARRC decided to let them and the Community Church erect buildings on Corporation land in the community center.

I think that most of the credit for the board's decision to allow churches on Corporation land could go to Dr. Albrecht. He worked doggedly and tirelessly to get at least one church there. He never felt it was right to have a community center built minus a church building. Of course he had lots of help and others carried their share of the responsibilities, but all would have given up long ago and built elsewhere, except for his persistence. We certainly were fortunate to have him in our community.

The board also approved the sale of equipment, livestock, and farm improvements to all farmers in the vicinity of the project, including settlers. The only requirement was that settlers join the project's cooperative marketing association. This action formalized the policy already begun the preceding fall.[8]

Ross Sheely reported at the meeting that 155 families were left in the project, 55 families had left, and 10 families were added to colony from local applicants. The board desired that 15 colonists be selected for vacant tracts. It decided six vacancies would be filled by Washington, D.C., and nine by the

ARRC Administration Building

ARRC in Alaska. Luther Hess, board vice-president, felt that the number of colonists could be brought back up to 200 and the board agreed. The board also felt that new colonists should not be brought to the project until 1937, when farm tracts would be ready for them. Barns needed to be built on many tracts, and land clearing was also required.

FERA administrators in Washington, D.C. evidently disagreed, since they began sending replacement colonists to Palmer that summer. While the ARRC wanted to bring the number of occupied farms backup to 200, the project's roles never exceeded 170 colonists after 1935.[9] That the number of colonists never again reached 200 can probably be attributed to the fact that new colonists had to pay their own transportation to Alaska, reducing the number who could afford to come, and to the fact that the ARRC did not have enough funds to develop additional tracts.

The first school baccalaureate in Palmer was held Sunday, May 17. All the churches in town combined for the service and Father Sulzman gave the address. Commencement was May 20. It took very little time for the senior high graduating class

Birthday Celebration

ARRC hospital

to receive his diploma. Earl Barry was the sole graduate. Eighth graders also received diplomas.[10]

May 22 was the picnic for the Wasilla school. That Friday, our family had its first experience of being up in the foothills of the mountains. The Wasilla School had its picnic up at the mouth of the canyon that leads to Fishhook Inn, and Fern, and Lucky Shot Mines, in the Talkeetna Mountains (Hatcher Pass). The canyon is only about three miles from our place, but we had never been there before. It was beautiful. A very rapid stream, with snow on the edges, rushed down over boulders at the base of the canyon, and the nearly perpendicular rock sides of the canyon were covered with mosses and flowers and small shrubbery.

The next day, the 23rd, I went down to Palmer. To think, that exactly a year before we had pulled into Palmer for the first time—so hungry, and so tired, and so anxious to get into our own place, cook a cup of coffee our way, and sleep in our own beds. We had to be satisfied with cheese sandwiches, and condensed milk in muddy coffee, and sleeping in the train in improvised beds with no pillows or blankets. When we looked out the train windows to the west we saw tents, people, and confusion, a dingy box car depot, and one tiny general store. Now, to the east we saw only woods and one farm house.

There were still tent houses to the west of the train tracks, but in neat rows back of the warehouse tents, press office tent, engineer's tent office, and tarpaper timekeeper's office. There were three restaurants, a hotel, a combination "photograph-souvenir-barbershop" building, a quite completely stocked general store, a drug store, and several small cabin residences, all of frame. There was a sign out for a lawyer and an insurance agent's headquarters.

Where the colonists' tents used to stand in Palmer there was now a newly seeded grain field. It really was not what we could call beautiful, but at least it showed growth and advancement.

Then to the east! A lovely depot (one a larger town than ours could be proud of) greeted the arriving. Over beyond that were all of the new Corporation buildings, all painted cream and tan and set behind the park area.

There was the pump station and new water tank, the garage, the power plant, post office, warehouse, blacksmith shop, cabinet and woodworking shop, shoe repair shop, barber shop

ARRC Cannery and Creamery

Birthday Celebration

and women's shop, the trading post with office and two wings, the big community hall and three story schoolhouse, the hospital and the dormitory, and six official residences. The creamery and cannery would be built that summer. Men were at work leveling and landscaping and surveying for streets.

Believe me! There was plenty of activity in 1936, with cars of every make and vintage, trucks, tractors, machinery, and people everywhere—on horseback, on bicycles, and afoot with packboards. One just planned to spend most of a day if he went to town, because every place was so busy we always had to stand in line. Then too, there were so many people to stop to visit with.

Looking at Palmer that day, I was proud of the small, but I think important, part that I played in the city's growth during that first year in the Matanuska Valley. We could laugh about some of the difficulties of that first year—like getting roast beef for weeks and weeks from the commissary, no matter what we ordered. When we asked if they could give us a break, since other colonists were getting pork, they sent us pork for weeks at a time.

But the experience wouldn't have been nearly as interesting if everything had run smoothly. Even carrying water from the spring, while it seemed a bit annoying, got pretty cold some-

ARRC Commissary/Trading Post

Market Street in the spring

times, and came as close to being a hardship as anything did. It looked not so bad in retrospect. In fact, we rather remembered with fondness those trips together, just Neil and I, on those gorgeous moonlit nights.

On that May day in 1936, I looked with eagerness at the next year and the years that would follow. From our hilltop home, I could look over our farm, at the Matanuska Valley, the Talkeetna Mountains, Knik Glacier, and the Chugach Mountains.

Like many other colonists and settlers, I had conquered the valley and its surrounding mountains, at least in my mind, and made them my own.

My over-riding ambition during that first year had been to tough it out and become a sourdough. I had accomplished that. Now I could look back with amusement at the troubles of the first year and wear with pride the "plaid shirt of the sourdough uniform."

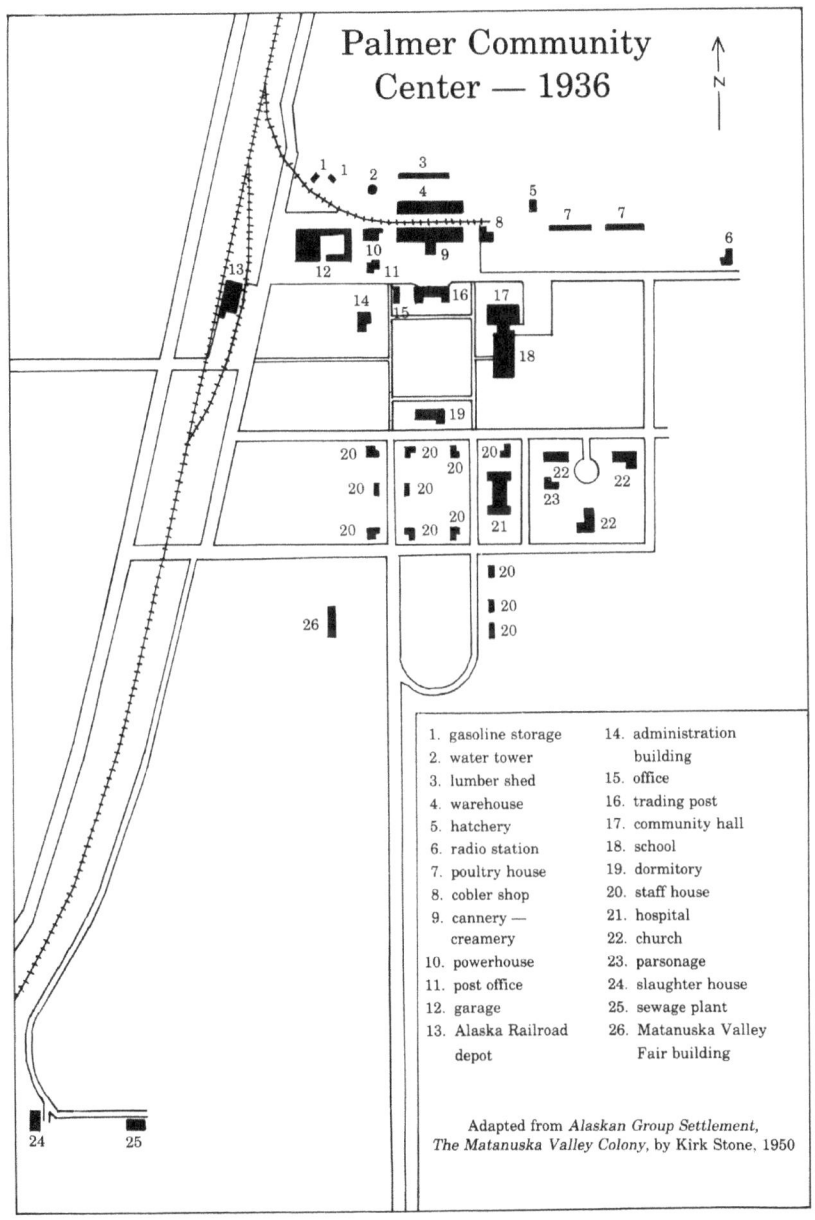

Thirteen
Triumphs and Tragedies: 1936-1937

Mr. Sheely took us up into the foothills of the Talkeetnas—towards the gold mines. Mr. Shonbeck (an Anchorage businessman with a large farm in the valley) has 300 cows with calves grazing up there above timber line. There are miles and miles and acres and acres of good grazing ground there. Sheely's used to have a large ranch in Arizona, and the white-faced cattle and miles of grazing land was like getting home for them... The scenery along the way was marvelous—up through the canyon and on up toward the top of the mountains—with glacial streams falling down the precipices of the canyon and the rock jutting out on the mountain sides—and the miles and miles and miles of rolling hills—and valleys with mountains in the distance. It seems to me that I've lived more thrills and experiences in the past year than I ever did in my whole life before.

The rush to complete our houses delayed clearing considerably, and we found clearing much slower than anticipated. Some colonists, like us, cleared land by hand, a tedious, time-consuming process. Other colonists were able to use Corporation tractors for clearing, but found the tractors too light for most clearing work. (The tractors were replaced the next year with heavier models.[1])

Don Irwin resigned as assistant manger of the ARRC in June to return to his old job as superintendent of the Matanuska Experiment Station. He was replaced by Leo Jacobs, who came to Alaska with the first contingent of colonists in 1935, and worked as the project architect. (He designed the Palmer Community Church.)[2]

We completed most of our planting by early June. Harold Johnson and Neil leased the cleared land on the vacant tracts in Camp 7 and were able to plant 15 acres in hay. Neil also began work on a building over our well to house the well pump, a workshop, and laundry room. Some colonists, in addition to planting in June, harvested wild hay from the flats below Wasilla, near Knik Arm. The cured hay was to be baled by the ARRC and trucked to the farms of those involved in the harvest.³

Wild hay harvested on the Hay Flats near Wasilla. Note the men with measuring tape at the front and back of the foreground stack.

Sheep, a draft horse, goats, and chickens arrived at our farm in July. We also decided that we needed more dependable transportation than the old Ford, so we bought a new GMC—a 1 1/2 ton truck. We bought the truck on the installment plan, as many other colonists in the valley did. The *Valley Pioneer* wrote that there were 60 new automobiles in the valley that summer.⁴ Also, anyone having a private vehicle was required to have a drivers license by June 15.

The Fourth of July brought Palmer's first Independence Day celebration. I was not too impressed with the festivities. There was a parade and I was one of the parade judges. Neil marched with the American Legion, and Mardie and Priscilla rode on the 4-H float. Later in the evening there was a free movie in town and then a dance.

One of the most anticipated events at the celebration was a baseball game between teams fielded by the Lucky Shot Mine (in the Talkeetna Mountains) and Palmer's Colonial Fire Department. During the lead-up to May's Colony Days celebration, the

Baseball game in Palmer—summer 1936

miners at Lucky Shot had challenged the men of the fire department to a baseball game to be held on the Fourth of July.

At that time there was no ball field in Palmer, so the next month was spent furiously developing a field next to the community center. With the field completed by early June, men from the fire department practiced diligently, and had their first game on June 14, resulting in a 4-4 tie with the Anchorage ball club.

During the July Fourth game the men from the Lucky Shot took a drubbing with a final score of 10-2. The *Valley Pioneer* reported that the miners took their defeat in good spirit. The paper also forgave them for their poor showing, explaining that since the mine's team was composed of men from both its day and night shifts, it had been impossible for all the team's members to practice together.[5]

July's most thrilling event for us was the connection of our bathtub to the cesspool. Finally, we could take real baths. I soaked in the bath every day for a week after the tub was in! Mardie and Priscilla each had friends come to stay overnight, and one of the main features of the visits was the tub bath.

A bulldozer arrived in late July to begin excavation for our barn. Building material for the building also began arriving.

We started building the barn in August and completed it by October. A tractor also arrived to plow and pull stumps on the land cleared during winter and spring. By the end of July we had several additional acres cleared and plowed and ready for discing.

In a way it was too bad we lived so far from Palmer, since our family and others in the outlying camps often had to wait what seemed unreasonable periods for equipment to arrive. The camps nearer Palmer got their fields plowed and disked first. Our neighbors, the Raschkes, quit the colony in August because Gus was so disgusted with the equipment delays.

Construction of the three churches in the community center also started during the summer. The Catholics built a frame structure and were the farthest along by August. The Lutherans bought several old government buildings at Eska, a camp along the railroad towards Chickaloon, and used the materials from them to build their church. We constructed the Community Church of 1,000 locally-harvested logs.[6] (All three churches were completed in 1937.)

We began harvesting hay in August, 1936. The hay, a combination of field peas, oats, and vetch, was high enough to reach the windows of our truck, about four feet tall. School began the first week in September in the new school building in Palmer. There were 271 pupils and 14 teachers. Neil was transferred from the Wasilla school to Palmer, and didn't have to drive bus, greatly simplifying our lives.

The first Matanuska Agricultural Fair was held September 4 through 7 at Palmer. It surely was a fine fair for our first attempt. There were very nice displays in all categories, although the livestock department could have had more entries. But it was so hard for many to get back and forth twice a day to take care of the stock, so they didn't bring much in.

The garden and canning exhibits were most wonderful though. One of the homemakers had a booth with 150 cans of Matanuska Valley products—no two the same. In addition to the regular line of vegetables, pickles, berries and such, she also had

A Creek, a Hill, & a Forty

Stock barns under construction at the fair grounds.

cans of milk, cream, cheese, moose meat, ptarmigan, and spruce hen. So you could see that with some ambition, there was no reason we couldn't have a varied diet year-round.

We entered about anything we had on the farm in the fair, and took first, second or third prizes for our goats, chickens, pig, heifer, baked goods, and quilt. All told, we collected $47.50 in premiums.

Neil was the assistant superintendent for the children's department, and I was the assistant in the handicraft department, so all in all, the fair meant a lot to us. In addition, a ceremony at the Knik River bridge on Sunday, the first day of the fair, officially opened the new highway between Anchorage and Palmer.[7]

Replacement families from the States began arriving in Palmer during the summer. New families moved onto vacant tracts in Camp 7 in September.

The ARRC exercised due diligence in selecting new families. All were required to fill out forms showing farming experience and other employment, family history, and references. Approximately 1/3 of the replacements were personally interviewed by FERA officials.[8]

Replacements from the States were required to have at least enough funds to transport them from their homes to Alaska. They took over tracts on the same conditions of credit and aid as the original colonists. Twenty-eight families left the project in

Highway bridge over Knik River—1936

Palmer to Anchorage highway between Bodenburg Butte and Knik River.

1936, and they were replaced by an equal number of new families, keeping the number of families in the project at 164.[9]

Hunting season opened in September—the first legal season for us. It seemed that every man was out hunting moose, bear, rabbit, grouse, and any wild game they could get near. Our life was so full of guns, ammunition, birds, hunting talk, and hunting plans, that we scarcely could get room for anything else.

A Creek, a Hill, & a Forty

But Masonic clubs, Legion and Auxiliary, Homemakers, church Council, Sunday School board, choir practices, 4H club work, Grange, etc. still presented themselves and had to be dealt with in turn. We also squeezed in a little housework on the fly, in addition to caring for all our livestock.

Neil did not even get a moose, as hard as he tried, but I think over 30 moose were killed in the area in September. We did share some of our neighbors' fortunes—at least one time we thought we had. Neil hired a neighbor's son to help clear land, and when fall came, the boy quit to hunt moose. Several days later, as I returned to the house from watering livestock, I saw the boy leaving the driveway and found a saddle of meat on the back porch.

Assuming the roast was from a moose the boy had shot, I soaked it in vinegar to reduce the wild flavor. Then I invited people from the University Extension Service, who were in Palmer for a meeting, out to the house for dinner. My guests raved about the tenderness and mild flavor of the moose. It wasn't until several days later, when talking to the boy's mother, that I discovered the roast was from the neighbor's old Holstein milk cow!

The Matanuska Valley Farmers Cooperative Association (MVFCA) was incorporated on October 28. and Neil was appointed to the first MVFCA board of directors by the colony council. (The organization's name was later changed to the Matanuska Valley Farmers Cooperating Association to comply with Alaska's territorial laws[10])

From the Matanuska Colony's inception, the ARRC had planned on a farmers' cooperative taking over ownership and operation of its agricultural facilities. In fact, producers' cooperatives were at the heart of most FERA rural rehabilitation projects across the nation. The cooperative element of the Matanuska project galled numerous early critics, and they likened cooperatives to communist collectives, stripping individual colonists of their freedom.

The MVFCA was to gradually acquire facilities from the ARRC. Until it assumed ownership, MVFCA supervised the

Triumps and Tragedies - 1936-1937

profitable facilities of the project while the ARRC retained the responsibilities for the facilities' operations. The ARRC also retained supervision and operation of the money-losing facilities in the project.

That meant that the Corporation paid for the operation of all facilities, including the necessary but unprofitable ones, such as the power plant, while the MVFCA received all the profits from the prosperous ones. The MVFCA was to pay out half of the profits as dividends to its members, and build up a fund from the other half of the profits to buy facilities from ARRC.[11]

In November the steamship companies were tied up by a seamen's strike. Neil's pay checks from Juneau were up to two months late because of the strike. The strike meant no freight for the Alaska Railroad, and without revenues the railroad had to lay off some of its employees. The ARRC was also low on funds and most of the construction workers were laid off. A few of the barns under construction were not completed when the ARRC workers were discharged, and colonists were left to complete the barns themselves or wait until 1937 for ARRC assistance.

By the winter of 1936/37 the Matanuska colony was beginning to move beyond its construction phase and the Palmer area was taking on at least an appearance of normalcy. Articles in the *Valley Pioneer* covering cosnstruction of ARRC buildings, personnel comings and goings and such were rapidly being replaced by news of weddings, births, club happenings, and school events; as well as advertising for local businesses, and a small classified section.

Jack Allman published an editorial in the November 19 issue of the *Valley Pioneer* entitled, "Palmer's Boom Days are Over." In it he wrote, "When the blizzard hit camp last Saturday bringing the season's first snow the wind screeched gleefully around bunk tents and covered the mud-frozen footprints under a six-inch mantle.

"Old Boreas [the Greek god of the north wind and winter] might have enlivened his spirit with the thought that he was driving the tent-dwellers away, but he was wrong if he did. It was the long-promised winter lay-off. Only skeleton crews remain, and barring the construction of a few barns and the digging of a sewage ditch, the boom days of the past two summer are over.

"From now on the valley will tend towards a slower, but steady, growth. New settlers will come in and farm some of the patented land surrounding the colonist tracts. Fingers of roads will reach out into areas open for homesteading and the forest will give way to clearings in which happy children will shout and play. The Matanuska Valley will grow as a community of farming homes. It will grow slowly, perhaps, but surely. Some of the present business houses may have a hard time hanging on, waiting for that growth, but it will come."[12]

One has to wonder whether, when Allman wrote that editorial in November, he knew (or at least had an inkling) that a month later he would be publishing the last issue of the *Valley Pioneer*.[13]

The construction workers who had bought the paper were gone. With waning interest from the general public about the colony, Allman expected subscriptions from outside the Matanuska Valley to fall.

The small population of the valley could not support a newspaper, and neither could the advertising dollars from the handful of businesses in Palmer. So the newspaper suspended publication. Allman hoped that conditions might improve by spring and he could resume publication, but the papers press was forever silenced. However, as Allman predicted, life in the valley went on.

By the end of 1936, we had completed our home's interior. We finally hired a carpenter to finish the interior work. Our barn was finished, and we had several outbuildings, including a well house and chicken house. We owned one cow and heifer, one horse, 26 sheep, 56 chickens, 1 pig, a dog, and four cats. Arabella added another heifer calf to our herd January 21, 1937.

Triumps and Tragedies - 1936-1937

My nearest neighbor and dearest friend, Fanny Johnson, died suddenly of a stroke in January, 1937. She had a stroke once before—the first summer we were here—when she was seven months pregnant.

They had to take the baby to save her, and the doctors told her she mustn't have another child. We found out that she was pregnant again when she died, but the doctor said that had nothing to do with her death. Some of us who were near her wondered if she were pregnant—but that surely she mustn't be under the circumstances.

It was such a blow for all of us in Camp 7. Harold and his five-year-old son Jackie stayed at the house alone for a short while after Fannie's death, but Harold confided to us that he couldn't stand the loneliness, so some friends stayed with them after that. Harold felt he couldn't look after Jackie all by himself, so he took him to stay with relatives Outside.

He was adamant that he intended to stay in the project, but Harold took all of his valuable personal property with him, so that just in case he decided to quit the Colony, the rest could be turned over to the Corporation. He was a hard worker, but needed someone to take the lead. He was rather helpless without Fanny. She always did all his thinking and made all the decisions and carried on the business end of things. We never saw Harold again, and the ARRC moved a new family onto the vacant tract, a family named Wilson.

A "work/credit system" was initiated February 1, 1937, replacing the bingle system, which had not worked as well as anticipated. The new system was designed to stimulate work on tracts and provide colonists with a cash income. Under it, colonists were paid cash for clearing and other work on their tracts. Project officials surveyed each tract and determined the work accomplished, and work still to be done. They then discussed the situation with the colonists and decided how much would be paid for the work.[14]

The cash payments were loans to the colonists and were charged against their accounts. We wanted to keep our debt to a minimum, so didn't participate in the program, although we still cleared our land. The ARRC expected to get most of the money from the payments back through sales at the commissary, where colonists now had to pay cash. We still received credit for feed, livestock, seed, building supplies, and equipment.

The United Protestant Church, the "Church of 1,000 Trees," after completion in March 1937

Having to pay cash at the commissary seemed to have a restricting effect on many of the colonists' purchases. We noticed a much better judgment shown in the buying then, and commissary business dropped about 33 percent during February. The work/credit system also caused a swift change in the area's landscape. Trees were quickly toppled and land cleared so we could hardly recognize the road to town. It was too bad they didn't start the work/credit system the first year. (The system was used until the end of 1937, and during that time 940 acres of cleared land and 801 of slashed land were added to the project.[15]

In March 1937, I finally acknowledged to myself and the world that I was pregnant again. At the age of 39, and with my youngest child 12 years old, pregnancy meant quite an adjustment for me. I felt so self-conscious—there is definitely a difference between a "sweet young thing" having her first baby, and an old grey-haired woman having her last. I had also been too independent for too many years to really enjoy sitting back and not taking part in everything, and I rather resented having to check my speed. But I was happy, none-the-less.

The Vojtas moved on to the Raschke place. They were lovely people and we liked them so much. But he was a bad drinker. He would go on periodic drunks that lasted for days—and she was afraid of him when he was drunk. She had been very unhappy and homesick ever since they had been on work /credit, where he had a chance to have cash to spend, because it had meant more and worse drunks. This past week he went again—and she packed her trunks and left for Minneapolis. I don't blame her. None of the neighbors do—even though they like him when he is sober. But she has worked hard and tried to do the right thing—and he promises "never again." I don't think he can manage alone—so no doubt he will have to withdraw.

On April 26, our neighbor to the south of us—Marty Novak—fell dead in the barn of a cerebral catastrophe. It was their first wedding anniversary. He was such a fine hard working earnest kid and he and Dorothy, his wife, had been so happy together in their plans for their new home. He was just a young fellow about 22 or 23—Dorothy is just 20. Both were very popular and very well liked everywhere and everyone especially admired him for his grit and industrious ways. He was just a little fellow—shorter than I am and very slender; but never a shirker. He always carried his share of the load in spite of his size.

We can't help but feel "Fate" got mixed in her intentions. In Vojta's case death would have been kinder than what they now face—and here's Dorothy left without happiness too. I imagine Dorothy will give up her place, she couldn't possibly carry on alone. Fortunately they had no children, neither did the Vojtas.

Planting began again in the spring, and again, we in Camp 7 waited for plows and disks while tracts nearer Palmer were cultivated. We were only able to get the five acres we had plowed in 1936 planted in June, 1937.

A Creek, a Hill, & a Forty

Plows finally arrived in camp at the end of June. The plow ran from 8 a.m. to 9:30 p.m., followed by the disc from 9:30 p.m. to 9:30 a.m. Neil had to follow the tractor, throwing rocks and roots out of the way, and he worked 13 to 15 hours each day. The ARRC crews plowed about 14 or 15 acres that summer, bringing our total cleared land to about 18 acres. That was all the tillable land on our 40-acre tract.

Ross Sheely resigned as manager of the ARRC July 1. He said the reason was to devote more time to his own farm, which he had purchased from a settler, but I think he was probably tired of the bureaucratic hassles of the project as well. As with Don Irwin, we retained our friendship with Ross after his resignation. Leo Jacobs, ARRC assistant manager, was appointed the new manager. I had every respect and a great deal of admiration for Mr. Jacobs, but I don't believe he had much actual "power." Just like Sheely's situation.

(The unspoken story Margaret did not share about Sheely ,and which she was probably aware of, was the drama of his protracted attempts to rein-in a recalcitrant colonist, and when those attempts failed, to have the colonist evicted. Among other objectionable actions, the colonist, Charles Ruddell, had received dairy cows from the ARRC paid for on credit, had fed those cows primarily with grain bought from the ARRC on credit, and then sold the milk to customers for cash—pocketing the cash and refusing to pay anything on his ARRC account.)

Although Sheely's actions had the blessing of ARRC officials in Alaska, officials in Washington D.C., wanting to avoid any negative publicity, pressured Sheely into reinstating the colonist. Ruddell withdrew from the project by the end of 1937, anyway[16]

The actions of Ruddell point to a persistent problem with the colony—that colonists had little or no access to cash they could spend as they pleased. Many colonists lifted from the relief rolls in the home states had not been full-time, or even part-time farmers, and they were enamored to working jobs that paid cash.

Triumps and Tragedies - 1936-1937

After arriving in Alaska, some colonists showed little interest in working their farms. The thought of devoting all their energies to improving their farm tracts and living on credit from the ARRC until their farms became productive was perhaps intolerable, as was the thought of not having coins jingling in their pockets.

Our son, Timothy Miller, was born at the Palmer Hospital on September 12, 1937, after what seemed an eternity to me. He had red hair and weighed 9 pounds, 15 ounces. I had been sure the baby would be a girl since we already had three daughters, but I was very relieved it was a boy. Neil finally had a son! Tim was the 100th colonist baby born in the valley, and the 125th baby born in Palmer's hospital.

It looks as though another neighbor would soon be pulling pins. Sandbergs, who moved on to the Greene place are having domestic difficulties—due to the drinking on the part of the husband. She won't let him have any money for fear he'll spend it on booze and he claims he can't even get enough to buy feed. She was to leave on this week's (September 28) boat for the states—to be gone indefinitely. So I suppose it won't be long until he quits the colony.

Our new furniture also arrived in September, just in time. Neil simply did not have time to make furniture, develop the farm, and teach school. He drove to Anchorage one weekend and bought a table and chairs for the dining room; a davenport, easy chair, and rocking chair for the living room, and bedroom furniture for our room.

Oh, what a relief to sit on real chairs and eat at a real table! and to have our springs and mattress on a real bedstead instead of benches. The bed did not really feel any different when we were on it, but it surely was an ease to clean around, and made the bedroom look decent.

Several colonists announced in fall 1937 that they no longer required assistance from the Corporation through the work/credit program. They paid cash at the commissary, but still used Corporation credit for feed, seed, and equipment.

Colonists' debts continued to increase and were a major problem. The average debt by the summer of 1937 was over $12,000 and some colonists owed up to $18,000.

Both colonists and ARRC officials realized that no farm could support the debts already incurred, and that debts could not be allowed to increase. Project officials knew that closing the project was politically impossible, although many of us feared that would happen. A debt adjustment had been discussed earlier in 1937, and it was finally approved at the August 17, 1937 ARRC board meeting.[17]

The ARRC board of directors voted to lower colonists' debts by cancelling all medical, dental, and hospital charges; well-drilling charges over $25; livestock feed up to $250; the first purchase of clothing by colonists; and subsistence and clothing bills within the recommended budgetary allowances charged between June 1, 1935 and January 31, 1937. In addition, the cost of building materials used on colonists' tracts was reduced by $500. Allowances were also made for work performed by colonists on our own houses and land.

After these reductions were made, a further reduction of 20 percent was approved to reduce all debt to a manageable level. The debt reduction process was not speedy. A committee composed of H. M. Colvin (special council to the WPA), Leo Jacobs, and Ross Sheely worked on debt reduction. They agreed that a debt level of $5,000 was manageable by the colonists. Colonists were called in to discuss their debts, often several times, before a satisfactory reduction was agreed upon.

One of our neighbors told us that even after all the allowances and deductions, he still owed $11,000, so the Corporation cut that amount in half, and he signed up to pay $5,500. Our entire indebtedness didn't even reach $5,000! If only they could have cut our debt in half. Most of us had our debts reduced by spring of 1938.

The work/credit program was replaced in fall 1937 with the "security and development program." The ARRC wanted 15

Triumps and Tragedies - 1936-1937

tillable acres cleared on each tract, and the new program was primarily aimed at clearing land. Clearing payments were increased, and payments for most other work stopped. (Work under the security and development program continued until January, 1939. By September 1, 1938, there were 133 tracts with at least 15 acres cleared, and the total cleared acreage in the project was 2,648.[18]

We were forced in October 1937 to trade our pick-up in for a new one. The roads in the valley, while graveled, were extremely rough, and chewed up vehicles. Our new truck was heavier, with more ground clearance.

We also had a late harvest, after school had started, so Neil had to hire-out for most of the harvesting, an expensive proposition. We had seven dairy cows (a sizable herd by valley standards) by the end of 1937, so we had to expand our barn. The 32' by 32' barns built by the Corporation were too small to support any sizable herd, and the cost of expanding them so dairy herds could be raised was a major expense for many of us colonists.

Colonists still left the project and were replaced by new families. Sixteen families left during 1937, and 24 families were added to the project. There were 170 families occupying farms in the project at the end of 1937.[19]

By the end of 1937 all the ARRC's facilities had been completed. Facilities included the trading post, warehouses, garage, post office, power and heating plant, creamery-cannery, laundry, inn, cabinet shop, blacksmith shop, chicken hatchery, 13 project dwellings, barbershop, cobbler shop, administration building, community hall, school, hospital, and several fair buildings.[20]

In December, Palmer also witnessed the birth of a new newspaper, The *Valley Settler*, published by the MVFCA. The paper, actually a mimeographed newsletter, filled the void left when Jack Allman's *Matanuska Valley Pioneer*, stopped publication. The *Settler* published co-op related articles, as well as community news, announcements, and advertisements. (It was published at least through 1959.)

Fourteen

The Rest of the Story: 1938–Onward

According to the January 6, 1938 issue of the *Valley Settler*, out of 171 colonist families, about 70 individuals had jobs away from their farms, including 55 who were employed by the ARRC.

The corporation experienced a persistent problem with colonists taking jobs away from their farms. While many colonists felt outside employment was necessary to bring in extra cash to support farm development, others preferred cash jobs to working their farms.

The corporation tried to ameliorate the issue in various ways. At first it actively discouraged colonists from seeking employment away from their farms. Then when it realized the futility of that approach, it grudgingly allowed it. Unfortunately, there were not enough corporation jobs to go around, and colonists who landed jobs with the ARRC were envied, even accused of being corporation sycophants. The only way the situation could have been avoided was if there was full employment for every colonist, which was impossible.

The ARRC tried to lessen the problem by discouraging non-colonists from applying for jobs with the corporation. Then it instituted a rotating schedule for unskilled employees so more colonists could be hired. Finally, it converted many ARRC jobs from full-time to part-time so more people could be employed.[1]

A sheep growers association was formed in February [1938], and Neil was elected president. I felt at times that we took on too many social responsibilities, but Neil chided me. He said that Palmer was the community our children would grow up in and it was our responsibility to see that it developed properly, and that meant participation.

The MVFCA met in February to elect board members. (The first board, elected in the fall of 1937 by the civic association, had only been temporary until regular elections could be held.) Only one member of each family is entitled to be a shareholder, regardless of how much stock [farm animals] the family may own. And only those farmers (either settlers or colonists) who own stock are entitled to vote. Quite a few people did not buy stock. The rest of us feel that the success of the colony depends upon the success of the Co-op Assoc. It now controls the trading post, garage, filling station, and warehouse, so it really affects the whole buying and selling operations of the farmers. Many of the farmers who do not own stock are doing what they can to discredit the association and claim they are going to crash the election, and pad the vote to get men in different from the present ones.

    School was closed for the day so that young children could be left with older ones, and, others could attend the meeting. Up until the meeting, only men stock holders could vote, so women went to ask for the right to vote. The voting lasted until very late, and Neil was the only member from the previous board to be re-elected, and then just barely. Neil was satisfied that there were five very good men on board, when we decided to come home, so it wasn't until later when our neighbors, the Wilsons, came by on the way home, that we learned that Neil had also been elected. After the election was over, Mrs. Sandvik kept at it until she got the men (that were left) to consider an amendment allowing wives to vote. The amendment was tabled for further consideration. Her aim was to get it so not only men and wives could hold stock and vote, but also all the children! She didn't get the children in—so apparently there were still some "men" left at the meeting. The co-op made 3% profit since July (trading post, warehouse (feeds) and garage). Our share of the profit amounted to $25.

    In April, we took a walk around the 60 acres that belong to the place that used to be the Rasckes. The ARRC is considering adding that land to ours, as three families have tried now and been dissatisfied to stay on the place, so it has been condemned. But the place has much more level land than ours, and we will be very glad to have it added to ours. We hope we can get the building too. I really believe the dissatisfaction of all three families was something within themselves, not anything that had to do with the place. Of the nine original families of camp seven, we are the only family left. The rest have all withdrawn and new ones taken their places.

Margaret and Tim sitting in the Miller's truck (c 1938).

It's a peculiar fact that throughout the entire colony, the ones who asked for and received the most help, money, provisions, and all, were the ones to get tired of it and leave first. It is disgusting to think of the thousands and thousands of dollars spent on those people who showed no gratitude, grabbed everything they could, wasted unnecessarily then went back to live on relief in the States.

Now for the most part, the ones who are left are the ones who are grateful for the opportunity—have worked hard and asked little. Now the money is gone, and by September the colony will have to be almost entirely self-supporting.

I just believe we'll "show them" that we can be! We may have some thin times, if we have to depend solely upon our farms, until they are all cleared and under cultivation. But many

The Rest of the Story: 1938–Onward

of us have some other means of making money to help us along. Many work in the community center. Many work for the road commission, some in the mines, some in the fisheries and canneries.

We continued working on the house.... We have about completed the painting inside the house and have two floors varnished too. We still have a few windows to paint and wardrobe doors, and inside of cupboards to do. Neil is painting the outside of buildings on days it doesn't rain, and is building fence when it's too wet to paint, and digging out the cellar when it's too wet to fence.

We had a beautiful trip up into the Chugach Mountains. Usually when we "go up to the mountains" we go north into the Talkeetnas. I didn't know there was even a road up into the Chugachs. We went with the Bixlers (settler neighbors).

At Eklutna (across the Matanuska and Knik Rivers from here) where they have the big Indian School, there is also a power plant. The water that furnishes the power comes from a lake up near the summit of the mountains above Eklutna...

The road [to the lake] winds around and up for 12 miles—just a narrow rutted wagon trail affair—but passable in dry weather... The lake is beautiful. It's way up at timberline, and huge. Some say it is 12 miles long—some say eight. It goes behind a part of the mountain from where the road ends, so we couldn't see all of it.

But it is beautiful—with a lovely shoreline of grey sand and rocks. The lake is glacial fed, and the mountain with it's glacial formations looms above it and reflects into the waters. (And we went without our cameras!!)

In August the ARRC board authorized the general manager to fill vacant tracts by leasing them to qualified applicants. The lessees could purchase the tracts after two years, with payments made during the term of the lease applied to the purchase price.[2] New applicants were not screened for need, and new arrivals were treated simply as farmers.

The garden is nice this year even though we planted late. I took one picking of Swiss Chard to the cannery. They can our produce on shares; all we need to do is cut it and haul it down. They wash it and can it and we get a percentage (30%.) We got 41 no.3 cans (4 1/2 cups per can) of chard.

On September 12, 1938, the Matanuska Valley Project was transferred from FERA (WPA) to the Department of the Interior (DOI). Leo Jacobs resigned as general manager to stay with the WPA, and Ross Sheely was again appointed manager.

The *Valley Settler*, in its September 16 issue, stated that while the project would technically be under the DOI's Division of Territories and Island Possessions, it would actually be administrated by the Alaska Railroad. At that time, the railroad was still owned and operated by the federal government under the jurisdiction of the DOI.[3]

There were immediate concerns among many colonists about railroad control over the project. One rumor began circulating that the headquarters for the ARRC would be moved to Anchorage where the railroad's offices were. ARRC officials denied any plans to move its headquarters, and the corporation's office remained in Palmer.[4]

We thought that with Sheely again as manager and the railroad in charge, the colony would be better off—more business-like. It certainly was more business-like. The administration had no thought for the individual.

Sheely stressed in an October *Valley Settler* article that, "Due to the transfer, the WPA is through, and in the future every effort will be made to place the colonists on a safe, and self-supporting basis. Business policies on a sound banking basis will be distinctly established. The farmers will be required to establish good credit with the accounting department before entering into further loans or indebtedness and must offer security for the same." Perhaps because of colonist concerns about railroad control over the project, Sheely also emphasized that the ARRC board of directors would "authorize all plans of administration."[5]

The Rest of the Story: 1938-Onward

Everyone was dissatisfied, even old sticks like us. Colonists left by the dozens….. I was afraid at the time that before too many months passed, there was going to be a crack in matters that would either end matters entirely, or give us a fighting chance. If Neil had not had a teaching job we could not possibly have stayed, as our family used the teaching salary to keep the farm going.

It would have felt terrible if anything had happened so we no longer had the community to live in. I enjoyed everything so much and we had such good friends. After one had worked as hard as we had and built up the kind of community we wanted, it became pretty much a part of us and it would have been exceedingly hard to leave.

The last week in November we had a terrific windstorm. The wind grew to almost hurricane proportions. Trees were down, windows of the chicken house blown out, etc. Some places the roofing blew off the buildings. One man had the whole end of his garage blown out. The fires long thought dead, in bulldozed windrows, were fanned to life and the men fought fires all week.

At Sandbergs (cornering us to the southwest) a spark from the chimney set a fire that ran through the woods for a long way. It grew so bad that they let it run through the woods, while all the men did was to protect Sandbergs and DeLand's buildings. There were serious fires all through the valley, but no buildings burned.

In keeping with the new unsympathetically-practical business attitude taken by the ARRC under the DOI, the corporation started eviction proceedings in November against four farmers who refused to join the MVFCA. Interestingly, the now more-business-minded ARRC bolstered its position by harkening back to the position of the ARRC under FERA administration, when social welfare was emphasized. In the November 4 issue of the *Valley Settler*, Sheely said that, "One of the controlling principles behind federal relief and rehabilitation projects was that the right of the single individual must be subordinated to the rights of the many"—essentially saying that unless all farmers worked through the farmers' cooperative, the project would fail. On a brighter note, the ARRC added a colonist, Frank Linn, to the board of directors that month. This was an action colonists had long been urging the ARRC to take.[6]

1938 was a discouraging year for many of us. Thirty-seven colonists and their families left the project by the end of the year, but only nine new families were added, reducing the number of colonists to 143.[7] Herrieds withdrew from the colony. When they first came here they were so enthusiastic and Leonard worked like a nailer. Ella worked hard too, and while she did a lot of beefing and complaining about the ARRC, I thought they'd stick. But before they'd been here two years, he had "washed-up." Wouldn't work, but just sat and whined. When credit was taken away and they had to work to pay for supplies, he still refused.

They had quite a time and finally they signed up to withdraw. They wanted their way paid back to Wisconsin, but the corporation wouldn't do that, giving Leonard a job to raise the money instead. He didn't work long, and finally the authorities decided it was easier and cheaper to let them go than stay. Then Leonard announced he had work elsewhere in Alaska, and Ella and the children would only be going back to the States for a "visit", but will return to Alaska.

At the January 1939 MVFCA meeting Neil declined nomination for re-election to the board of directors. He wanted to try for the coordinators job instead. They kept him on as the secretary-treasurer, though [with the understanding that Neil's duties would continue until the secretary-treasurer could be made a salaried position]. If he doesn't get the coordinator job, he will go outside to summer school. Also, at the meeting, the amendment tabled at the last meeting to allow more than one member of a family to vote was defeated.[8]

The MVFCA's annual report in the February 3 *Valley Settler* stressed that the relationship between the co-op and the ARRC was, at times, less than congenial, and that a permanent co-op manager was needed. In the *Settler's* next issue it was announced that Mr. Clair L. Stock, with 18 years of experience with co-op associations was given the manager position. Before coming to Palmer, Mr. Stock had been managing a co-op in Kelso, Washington.[9]

Beginning in early 1939 a group of dissident farmers began using the MVFCA to voice their opposition to ARRC policies and management. The core members of this faction called themselves the Matanuska Valley Protective Association. They also dubbed themselves the Order of Ice Worms. One of the farmers elected to the MVFCA board in January, a replacement colonist named Carl Rasmussen, became one of the Ice Worms' leaders.

The historian, O.W. Miller, wrote, "The Ice Worms discontent… was unspecified, broad, unfocused—except that the ARRC was the source of credit, supervision, and nagging guidance. In a midwestern farming community in a bad year, resentments would have been scattered to include local bankers and lawyers, the railroad, dealers in farm supplies, the representatives of several levels of government, and the solidly prosperous members of the community. In the Matanuska Valley the ARRC stood in place of all these.

"The resentments were too scattered to be put clearly or openly, but if a colonist worried about his present short-term debts or the land payments that would begin in 1940, if the reward for his efforts seemed always in the future while he had little present cash in his pocket, if he was envious of the salaried affluence of the ARRC management or the regular ARRC jobs held by some colonists, or if he felt painfully cramped in the range of decision open to him, he located the source of his discontent in the corporation and identified his interests with the cooperative."[10]

The ARRC started a new clearing program in January 1939 to increase the cleared acreage to 30 acres per tract, even though some tracts, like ours, did not have that much tillable land. The clearing rates were increased, and by July, a total of 3,577 acres had been cleared in the project.[11]

Two articles in February issues of the *Valley Settler* wrote of unsanctioned meetings sponsored by the Ice Worms during February. The first meeting, attended by about 35 people was held at Matanuska. The second, held at the Palmer Civic Center, had about 240 in attendance, including Neil. According to the *Settler*, after a report was given at the second meeting, there was a "lengthy discussion of conflicting opinions which resulted in a near riot."

After order was restored, Neil rose, and offered a motion condemning the actions of those responsible for the meeting, and accusing them of trying to usurp the authority of properly-elected officials. He also moved that the Civic Association "request the ARRC, the Territory of Alaska, and other civic and business enterprises refuse to recognize any but the duly constituted authorities of the Valley: the Civic Association and the MVFCA." The motion carried overwhelmingly and the meeting quickly ended. For the time being, the Ice Worms were hoisted by their own petard.

Strangely enough, a letter to the editor in the *Settler* (the same issue that described the second Ice-Worm-sponsored meeting) railed against the recently-elected school board for passing a resolution "That no farm-residing or owning teacher is to be retained…"—a resolution that seems squarely-aimed at Neil—and one that demonstrated the tensions within the colony.[12]

Mardie graduated from high school in May, 1939. She wanted to attend college Outside, but we did not have the funds to send her to school, so she stayed at home and took additional classes at the high school. She eventually landed a job at the railroad hotel at Curry, half way between Seward and Fairbanks. [The next year she transferred to the Mount McKinley National Park Hotel.]

By 1939 most colonists' farms were still not self-supporting, and few people could make their debt payments. The ARRC

The Rest of the Story: 1938-Onward

Neil and Tim doing chores (c 1939)

gave Ross Sheely the authority to accept partial payment from farmers making real progress on farms.[13] Ross resigned as general manager in October 1939, and was replaced by Dr. Herbert Hanson in January, 1940. Mr. Hanson was a stern administrator and was not popular with most of us. His views of how the project should be run often conflicted with those of the MVFCA, and with those of individual colonists, such as Neil.

The MVFCA negotiated with the ARRC during the winter of 1939-40 to assume ownership of the Corporation's processing facilities. It obtained a $200,000 non-interest bearing 30-year mortgage from the ARRC for the facilities, and received a $300,000 operating loan from the Farm Security Administration.

The facilities were transferred to the MVFCA in mid-January 1940. The selling price for the facilities was eventually set at $200,000, less than half the appraised value.[14]

The co-op meeting in February 1940 showed us just how strong the Ice Worm element has become. They had their men all elected. Every one of the four new co-op board members went in on 69 votes out of 120. The rumor is that they plan to put Stock out as Coop manager, and put Rasmussen [one of the leaders of the Ice Worms] in. But so far they've done nothing drastic.

The editor of the *Valley Settler*, Marie Wilson, who had helmed the newspaper since December 1937, refused to publish an article from the co-op board in January 1940 because she viewed it as libelous. She had been sometimes criticized for being too friendly towards the ARRC, and after the Ice Worms took control of the co-op board of directors she resigned as the newspaper's editor.[15]

Mr. Stock left for the Outside on co-op business Saturday morning (Feb. 17). That same day circulars were sent out to individuals throughout the community that were "petition heads" demanding the resignation of Mr. Stock as manager.

Marie Wilson got busy and traveled all over the valley notifying those she knew weren't 'Ice Worms' (supporters of Rasmussen), that we'd have a meeting that night to plan some sort of protective action. So the battle is on. (The effort to remove Stock as MVFCA manager was unsuccessful.)

The afternoon of February 25, we had another meeting of those who are known to be definitely opposed to the practices and policies of Rasmussen and his Ice Worms. This is just a group who meet to try to determine means of saving the co-op from being broken up by the radicals.

Our group grows with each meeting. Today we drew up a list of things for which we stood and we are going to ask each person in the co-op to sign it. It definitely opposes the things the Ice Worms are doing. So anyone who refuses to sign the resolutions we will know is an Ice Worm.

Then we plan a definite program of educating the community about cooperative lines, and hope to get some of the Ice Worms to see they are hurting themselves more than anyone else by such practices as the Ice Worms advocate.

The valley is definitely divided into two factions—the Ice Worms (many of whom are non-producers, and those who don't intend to pay bills and are definitely at war with the ARRC in everything), and the other faction... called the "Producer's Union," made up of those who have tried to develop their farms, pay their debts, and build up the cooperative market.

It's a fierce battle and waging on all sides. You don't know whom you are talking to. Neil and I don't go to anything when we think such things will be discussed as it wouldn't be fair for the kids in school for Neil to mix into the community quarrel. Of course our sympathies go with the producers.

One of the most contentious issues that frustrated many colonists, Ice Worms and members of the Producers Union included, was the realty contracts that had been presented to them in 1938 after all the debt reductions had taken effect. Colonists had numerous complaints about the contracts, the most serious being the restrictive deeds they would receive after paying off their mortgages.

The contracts were structured so that the project, in keeping with its *raison d'etre*, would essentially remain agricultural and cooperative in perpetuity. However, with a changing mood in Congress and the nation, the restrictions eventually disappeared from realty contracts, and colonists received clear title to their tracts when deeds were finally issued.[16]

Kleinpiers and Swaboda will both be gone for the summer. In June Neil goes to Snug Harbor to work in a cannery. Camp 7 promises to be manless this summer.

Mr. Hanson is a bureaucrat's bureaucrat and run-ins with Rasmussen and the coop make him just like a mad bull. Ross Sheely agreed to helped while Neil was gone. Ross felt that we wouldn't stay in the colony long and said he could foresee that someday soon we would get so mad at the whole thing, we'd just pull out. I told him we wouldn't do that, but I didn't think we'd do much about putting money in to the place to build it up, just to be kicked out and give the ARRC the benefit of it.

By 1940 the average cleared land on developed tracts was 23.5 acres, 1.5 acres over the amount planners in 1935 had anticipated would be cleared on each tract by 1940. It had taken much cajoling and enticement from the ARRC, but at least the goal for cleared land had been reached. There were 4,312 acres of cleared land in the project by August of that year.[17]

Few of our farms were profitable by 1940, and the markets for agricultural products were still poorly developed. Many of us depended on outside incomes to support our farms, and we were actually part-time farmers. The lure of high-paying salaries in Anchorage, constructing defense facilities, attracted some colonists, and the money from these jobs kept many farms solvent.

The first payments on our realty loans were due in 1940, but few of us were able to make payments. The ARRC board of directors authorized the general manager to adjust or postpone payments on debts for farmers making sincere efforts to develop their farms.[18]

After farmers settled down to paying their mortgages (some farmers very grudgingly), the valley quieted down from the bickering between the ARRC and the Ice Worms. The corporation was quickly ending its role as land developer, as idealistic community planner, and as paternalistic advisor. As the colony matured, and as the federal government's view of its role in an individual's affairs changed, the ARRC became primarily just a land owner and loan processor. Grumblings against the ARRC continued for years, but the open antagonism between the ARRC and the MVFCA was over. Within a few years the Ice Worms faded away.

Of course, we were trying to develop our farm and keep it solvent, so our debt payments were adjusted. We planted 34 acres in 1940, part of that acreage rented from a neighboring settler. Twenty-five acres were planted in hay, two in oats, two in barley, two in wheat, two in potatoes, and one for our garden. We had 1 horse, 10 sheep, 100 chickens, and 10 dairy cows.

Our first cow, Arabella, proved to be an exceptional animal. She was not a good looking dairy cow, but she was an excellent milker, and gave birth to seven heifer calves before we sold her. After being sold, Arabella only gave birth to bull calves.

Neil had not had any success back in Wisconsin getting his dairy cows to produce heifers. Our friends told the story that in Wisconsin, all he could get his cows to produce was boys, and his wife only girls, but after moving to Alaska, all he could get his cows to produce was girls, and his wife only a boy.

Priscilla graduated from high school in May, 1940, and worked at a cannery in Kenai for the summer. In the fall she moved to River Falls, Wisconsin to attend college. She stayed with her grandmother while attending school.

The nature of the farms in the project began to change around 1940. They had originally been planned as subsistence farms that would provide families with most of their food and generate some cash to provide for the families' other needs. But most farms gradually changed into specialized commercial enterprises such as dairy farms. Of the 118 tracts being farmed in 1940, there were 83 general farms, 9 dairy farms, 6 truck farms, 2 poultry farms, 1 sheep farm, and 17 unclassified farms.[19] Ours was one of the unclassified ones.

The ARRC had to deal with many farmers' inability to make debt payments again in 1941. The board of directors authorized deferring of realty loan payments two years for farmers in good standing.[20]

Even in 1936, we discovered it was very difficult for Neil to teach school in Palmer and farm at Camp 7. Commuting was too big a problem with the rough roads in the valley. We tried to trade our farm for one closer to town but were unable to. We also found it extremely difficult to make the farm profitable, and year after year, Neil's earnings from teaching supported the farm. We had increasing differences with Dr. Hanson, the ARRC manager, and with the manager of the MVFCA. In 1941 we sold our stock, relinquished our tract, and moved to Palmer, where Neil continued to teach school.

Construction at the new Army base in Anchorage, as well as civilian positions on post, provided more employment opportunities for colonists, but it also exacerbated the flight of labor away from valley farms. In 1941 there were so few colonists working their farms that there was a shortage of workers for the fall harvest. The Army Air Corps volunteered 40 enlisted men to help dig potatoes and carrots in return for a share of the harvest. The airman also helped with the hay harvest.[21]

While Mardie was working at Mount McKinley National Park, a young man named Jim Teale started work there. Jim had worked his way to Alaska on a steamship, and then landed a job at the park hotel. Mardie and Jim fell in love and after a courtship in the middle of the Alaska wilderness were married at Palmer in the fall of 1941.

At a dance in the spring of 1943 Janell met a young man named Jiggs Mickel, who was stationed in Anchorage with the Army Air Corps. They hit it off immediately. Janell graduated from high school in May and landed a job with the U.S. Signal Corps in Anchorage. Jiggs and Janell wanted to marry then, and even though Neil and I liked Jiggs, we felt Janell, now 17, was too young to marry.

Jiggs was transferred back to the States that summer, and Janell spent a miserable fall pining for him. She and her boss talked Neil and me into sending her to Pocatello, Idaho to visit Jiggs at Christmas. As soon as she arrived, Jiggs and Janell were married.

In 1946 Neil and I were able to buy a farm one mile north of Palmer. The farm had begun as Joseph Puhl's colony tract, but when Puhl quit the project, it was bought by Carl Wilson, the project's creamery manager. We bought the farm from Mr. Wilson. The house on the Puhl place was the first one completed in the colony. Puhl and three neighbors had tired of waiting for construction crews to build their houses, so they banded together and built their own, easily finishing before anyone else.

Priscilla married Dexter Bacon in 1946. She met him while attending college. Dexter, whose family lived in Mabel, Minnesota, was in the Navy at the time. After World War II

The Rest of the Story: 1938-Onward

The Puhl farm, which the Millers bought in 1946. The cabin was the first colonist home completed in 1935.

Neil mowing hay (late 1940s-early 1950s).

ended, Dexter and Priscilla were married at my mother's house in River Falls, Wisconsin. They moved to Alaska during their honeymoon, driving over the Alcan Highway soon after it was opened to civilian traffic.

Neil continued to teach and farm until 1955. Tim (18 years old) went Outside to join the U.S. Army Signal Corps that year. Neil accompanied Tim to the induction center, and while in the States, Neil decided that at the age of 59 he had seen enough

Alaska winters. We sold part of our farm to Priscilla and Dexter, and then moved to Arbon, Idaho where we both taught school. After retiring from teaching in 1964 we moved to Woodburn, Oregon.

Neil died in Woodburn on January 29, 1971, and Margaret died in Portland on March 17, 1989. Priscilla and her husband, Dexter Bacon, continued to live in the Puhl house in Palmer until moving into the Palmer Pioneer Home when their health started to fail.

Dexter died on January 22, 1999, and Priscilla died on Oct. 4, 2003. They raised four children in Palmer. Priscilla worked as a nurse at Valley Hospital, then as a public health nurse at the Palmer clinic and finally as a nursing supervisor for the Bethel area. She also served on the board of directors for Valley Hospital. After retiring from nursing, she partnered with Dexter in the family's building-supply store, Mat-Su Supply, working as bookkeeper and interior design consultant.

Janell Mickel passed away in Bella Vista, Arkansas on June 11, 2005. Her husband Eugene L. "Jiggs" Mickel passed away Sept. 29, 2013. Jiggs made a career of the U.S. Air Force, retiring in 1967 as chief master sergeant, having served in WWII, Korea and Vietnam. They had two children.

Mardi Teale died on Saturday, April 20, 2013. Her husband, Jim, died May 17, 2014. They had three children. During World War II, Mardi worked for the Civil Aeronautics Agency in radio communications. After graduating from the Herrington Institute of Interior Design, Mardi and her business partner formed the firm of Teale and Rudolph, Inc. in Chicago, Ill. She later moved to the West Coast, but remained involved with architectural design and the art world in California and Arizona.

Tim Miller passed away on January 14, 2023. Tim originally left Alaska in 1955 but came back several times, spending a combined total of 42 years there. Tim followed in his father's footsteps, serving in the army. After his military service Tim completed a degree in electrical engineering technology.

The Rest of the Story: 1938-Onward

Tim and his wife, Jean, raised four children while Tim worked as a telecommunications manager for the U.S. Fish and Wildlife Service in Idaho, Oregon, and Alaska. They moved to southern Utah in 2000, where Tim worked for the Bureau of Land Management.

Neil and Margaret's grandchildren and great grandchildren are scattered throughout the western United States, and one grandson, Ray Bacon, has retired to the Phillipines. Granddaughter Elizabeth (Betsy Bacon) Bonnell is the only grandchild residing in Alaska, albeit in Fairbanks. Her husband, Ray, with the blessing of Margaret Miller, edited and adapted her letters to form this book.

The Millers' first house in the Matanuska Valley, located along Palmer-Fishhook Road north of Palmer, is still part of an active farm. The farm the Millers moved to in 1946 has been subdivided, and the farmhouse there (which is on the National Register of Historic Places), is now within Palmer's city limits.

Appendix 1
List of Colonsits in 1935

Michigan

1. Anderson, Walter and Garnet, 4 children, Kenton
2. Bennett, William and Ruth, 4 children, Empire
3. Boice, Harold and Lona, 6 children, Merritt
4. Campbell, George and Onabelle, 2 children, Mio
5. Carter, Clifford and Dorothy, 2 children, Muskegan
6. Casler, William and Elsie, Mesick
7. Chaney, Escar and Louise, 3 children, Stephanson
8. Cousineau, Charles and Reba, 1 child, Lansing
9. Davis, Harold and Edith, 2 children, Lansing
10. Dingman, William and Mildred, 3 children, Frankfort
11. Dreghorn, LAwrence and Grace, 5 children, Wolverine
12. Durphy, Robert and Gladys, 3 children, Cheboygan
13. Ellsworth, Lester and Senia, 5 children, Merriweather
14. Ennis. Max and Lila, 3 children, Tower
15. Fitzpatrick, Theodore and Leonea, 3 children, Roscommon
16. Foster, Kenneth and Marion, 2 children, Stephenson
17. Fox, Waldo and Mable, 1 child, Halbert
18. Frank, Darrell and Lois, 2 children, Mio
19. Green, Clarence and Alida, Hancock
20. Havemeister, Arnold and Emmy, 1 child, Wallace
21. Higgenbotham, Robert and Clara, 2 children, Stambaugh
22. Hoeft, John and Adeline, 1 child, Rogers City
23. Hoganson, Harold and Mayme, 3 children, Ewen
24. Hipkins, Roy and Ada, 2 children, Arcadia
25. Huntley, Walter and Beatrice, 2 children, Sault St. Marie
26. Hynek, William and Neville, 5 children, Faithorn
27. Jacobson, Adolph and Mable. 2 children, Mohawk
28. Johnson, Arvid and Edith, 2 children, Crystal Falls
29. Johnson, Harold and Fannie, 1 child, Houghton

Appendices

30. Kalliosaari, John and Letta, 1 child, Copemish
31. Laako, Henry and Ruth, 1 child, Mohawk
32. Larsh, Emil and Gertrude, 2 children, Iron Mountains
33. LaValley, Edward and Florence, 4 children, Houghton
34. Lipke, Henry and Mary, 2 children, Harrietta
35. Loyer, Joseph and Naomi, 4 children, Harrisville
36. MacNevin Leon and Loraine, 5 children, Altanta
37. Manginen, Walter and Vivian, 1 child, Champion
38. Martin, Clyde and Arbutus, Mackinaw City
39. McCormick, Martin and Margaret, 3 children, East Tawas
40. Newville, Irving and Lila, 1 child, Boyne City
41. Nutilla, Eino and Betty, 2 children, Ironwood
42. Onkka, David and Saina, 6 children, Bruce Crossing
43. Pakonen, Russell and Madelon, Long River
44. Parks, Clarence and Beulah, 3 children, Merriweather
45. Parlette, Paul and Violet, Rapid River
46. Pfeiff, John and Clara, 6 children, Stephenson
47. Piskowski, Frank and Farnces, 5 children, Ironwood
48. Porter, John and Alice, Champion
49. Porterfield, Ernest and Almeda, Harrietta
50. Quarnstrom, Clarence and Sadie, 1 child, Daggett
51. Retallic, Arthur and Alice, 1 child, Amasa
52. Rotz, Fred and Emma, 2 children, Plymouth
53. Schutt, Claus and Hattie, 3 children, Lucas
54. Smith, Martin and Margaret, 7 children, Ewen
55. Snyder, Thomas and Freda, 4 children, Manistee
56. Spencer, Milan and Margaret, 1 child, Mesick
57. Spencer, Nelson and Olive, 3 children Chauncey
58. Stebbins, Dewayne and Evelyn, Mesick
59. Sturdy, Norris and Leona, 1 child, Crystal Falls
60. Sullivan, Carl Francis and Elizabeth, 1 child, Cheboygan
61. Usher, Robert and Helen, 3 children, Copemish
62. Van Wormer, Howard and Bernice, 1 child, South Boardman
63. Venne, George and Irene, 3 children, Bear Lake
64. Way, Sherman and Esther, 1 child, Lyon Manor

65. Wilding, John and Viola, 4 children, Cooks
66. Wilson, Amedee and Leah, 4 children, Cooks
67. Wirtanen, Elino and Fannie, Iron River
68. Zook, Harold and Clara, 3 children, Central Lake

Minnesota

1. Alexander, Harold and Frances, 4 children, Robbinsdale
2. Bell, Lloyd and Dorothy, Brook Park
3. Benseon, Henning and Iren, Barnum
4. Bergan, Leaonard and Alice, 1 child, Carlton
5. Carson, Arno;d and Hortense, 2 children, Milaca
6. Christianson, Martin and Leota, 1 child, Ogilvie
7. Cook, Clyde and Jesse, 7 children, Longville
8. Doughty, Glendon and Hulda, 2 children, Milaca
9. Eckert, Virgil and Lillian, 2 children, Cloquet
10. Emberg, George and Evelyn, 3 children, Proctor
11. Engebretson, Oscar and Johanna, 3 children, Richfield
12. France, Grant and Iva, 5 children, Grand Rapids
13. Fisher, Otto and Frances, 1 child, Kanabec
14. Fredericks, Albert and Audrey, 2 children, Sturgeon Lake
15. Fredericks, Allena and Lenore, 3 children, Sturgeon Lake
16. Giblin, Theodore and Jennie, 1 child, Brookston
17. Hack, Arthur and Mabel, 3 children, Ogilvie
18. Hemmer, Partick and Cora, 4 children, Wright
19. Hesse, Claude and Helen, 1 child, St. Louis Park
20. Holler, John and Gertrude, Pine City
21. Ising, William and Marie, 2 children, Saginaw
22. Jahr, Paul and Emma, 4 children, Pine City
23. Jenson, Henry and Edna, 2 children, Littlefork
24. Jones, Vernon and Eleanor, 2 children, Onamia
25. Johnson, Johan and Irene, 1 child, Anatol
26. Kenser, Grant and Gertrude, Swatara
27. Kerttula, Oscar and Elvi, 2 children, Deer River
28. Kindgren, Oscar and Saima, 1 child, Duluth

Appendices

29. Kirsh, John and Rose, 3 children, Solway
30. Kleinpier, Kenneth and Grace, 3 children, Anatol
31. Larson, Fred and Laura, 4 children, Ranier
32. Leander, Rudolph and Inez, 1 child, Onamia
33. Lee, Francis and Eugenie, 3 children, Forest Lake
34. Lemmon, George and Josephina, Littlefork
35. Lemmon, Gilford and Catherine, Littlefork
36. Lepak, Thomas and Irene, 1 child, Duluth
37. Lund, John and Margaret, 1 child, Saginaw
38. McKechnie, Loren and Edna, 5 children, Cloquet
39. Meehan, John and Viola, 5 children, Cloquet
40. Meier, Carl and Edith, 5 children, Duluth
41. Moses, Arthur and Blanche, 1 child, Bigfork
42. Moss, Edward and Myrtle, 1 child, Pine City
43. Olmstead, Vernon and Pearl, 3 children, Remer
44. Olson, Hilmer and Gwendolyn, 2 children, Duluth
45. Olson, Walter and Minnie, 1 child, Duluth
46. Patton, Claire and Margaret, 1 child, Long Siding
47. Peterson, Otto and Meryle, 5 children, Wrenshall
48. Pippel, Walter and Melva, 4 children, Robbinsdale
49. Poore, Cahney and Florence, 1 child, Littlefork
50. Powell, Ores and Lorenzie, 2 children, Duluth
51. Rebarchek, Raymond and Edma, 1 child, Graceton
52. Rorrison, Lawrence and Vera, 2 children, St. Louis Park
53. Rossiter, Henry and Mary, 4 children, Cloquet
54. Ruddell, Charles and Ada, 6 children, Duluth
55. Saaralla, Matt and Olga, 1 child, Swan River
56. Sieber, Joseph and Albertina, 5 children, Grasston
57. Sandvik, Ingolf and Agnes, 6 children, Moose Lake
58. Sjodin, Clarence and Alice, 2 children, Onamia
59. Smith, Lauren and Hollis, 5 children, Bennett
60. Splittgerber, Herman and Violet, 2 children, Hinkley
61. Stephan, Vincent and Betty, 1 child, Grasston
62. Strang, Eldred and Helen, 1 child, Strang
63. Swanda, Frank and Wilhelmina, 2 children, Pine City

64. Ubert, Lawrence and Alice, 2 children, Kingsdale
65. Vasanoja, Laurence and Helen, 5 children, Cloquet
66. Vickaryous, Anthony and Alys, 4 children, Bandette
67. Wilkes, Ray and Ruth, 5 children, Wahkon

Wisconsin

1. Anderson, Chris and Grace, 3 children, Shell Lake
2. Anderson, Clarence and Clara, 2 children, Draper
3. Archer, Pearle and Dorothy, 4 children, Nelma
4. Arndt, Lawrence and Etta, 1 child, Nelma
5. Bailey, Ferber and Ruth, 2 children, Lena
6. Barry, Earl and Louise, 7 children, Rhinelander
7. Beyland, Oscar and Irene, Rice Lake
8. Biller, Robert and Dorothy, 2 children, Fence
9. Bouwens, William and Lulubelle, 11 children, Rhinelander
10. Bradley, John and Sylvia, 2 children, Lake Nebagamon
11. Brown, Otis and Grace, 3 children, Enterprise
12. Campbell, Harry and Theodora, Abrams
13. Church, John and Julia, 5 children, Mountain
14. Connors, Goerge and Edith, 4 children, South Range
15. Covert, Albert and Catherine, Cable
16. Clayton, Walter and Rose, 3 children, Spooner
17. Dean, Ballard and Marion, 3 children, Fence
18. DeLand, Nile and Helen, 5 children, Winter
19. Dragseth, Joe and Velma, 2 children, Rice Lake
20. Ellison, Carl and Ruth, 1 child, Billings Park
21. Erickson, Carl and Inga, 3 children, Rhinelander
22. Ferguson, Walter and Mable, 2 children, Iron River
23. Greise, Raymond and Hazel, 3 children, Starks
24. Gulberg, Bernard and Beatrice, 2 children, Medford
25. Hamman, Leroy and Gretchen, 2 children, Iron River
26. Henry, Francis and Ella, 1 child, Viroqua
27. Hermon, John and Hilda, 6 children, Plymouth
28. Herried, Leonard and Ella, 2 children, Trempealeau

Appendices

29. Hess, Frank and Florence, 3 children, Cavour
30. Huseby, Einer and Inez, 3 children, Pembine
31. Jensen, Harry and Viola, 4 children, Tipler
32. Johnson, Clinton and Doris, 2 children, Winter
33. Johnston, Victor and Klaria, Harshaw
34. Johnson, James and Lillian, Shell Lake
35. Juvette, Eugene and Mary, Winter
36. Koenen, Henry and Bernice, 2 children, South Range
37. Kurtz, Cecil and Mary, 2 children, Suring
38. LaFlam, Claire and Emma, 3 children, Shell Lake
39. Lake, John and Magdalene, 3 children, Superior
40. LaRose, Henry and Clystia, 4 children, Phillips
41. Lentz, William and Viola, 1 child, Merrill
42. Lentz, Joseph and Zuleika, 6 children, Merrill
43. Mattson, Runar and Eleanor, Ashland
44. McKendry, Howard and Edith, 1 child, Winter
45. Miller, Neil and Margaret, 3 children, Blair
46. Monroe, Lester and Mary, 3 children, Hiles
47. Nelson, Arthur and Isabel, 2 children, Shell Lake
48. Nelson, Paul and Margaret, 1 child, Riverview
49. Nichols, Harry and Virgie, 1 child, Georgetown
50. Novak, George and Catherine, 1 child, Ashland
51. Puhl, Joseph and Blanch, 2 children, Rice Lake
52. Raschke, August and Anna, 1 child, Wentworth
53. Reitan, Bernard and Alice, 2 children, Superior
54. Ring, Frank and Lucille, 6 children, Beecher
55. Roughan, Henry and Minnie, 5 children, Monico
56. Runyan, Scottie and Irene, 2 children, Winter
57. Scheibl, Gustave and Alethea, 3 children, Sheboygan
58. Schultz, William and Martha, 1 child, Tomahawk
59. Sexton, Allan and Minnie, 5 children, Pelican Lake
60. Smith, William and Alice, 5 children, Bennett
61. Sorenson, Clarence and Vivian, 1 child, Rice Lake
62. Soyk, Martin and Genette, 1 child, Minocqua
63. Taylor, Neil and Violet, 1 child, Florence

64. Walport, Paul and Ethel, 1 child, Stone Lake
65. Weiler, Nicholas and Elsa, 4 children, Medford
66. Worden, Frank and Margaret, 5 children, Three Lakes
67. Yohn, Victor and Manila Bay, 3 children, Tomahawk

Oklahoma

Stahler, John and Elizabeth, 6 children, Chaney

Appendix 2

Matanuska Valley Settlement Agreement

THIS AGREEMENT made this_____ day of_____, 1935, between the ALASKA RURAL REHABILITATION CORPORATION, whose principal office is at Juneau, Alaska, hereinafter known as the Corporation, _____of the County of_____, State of_____, whose Post Office address is_____, hereinafter known as the Colonist, in behalf of himself and family, consisting of the following members:_____,

WITNESSETH, That

WHEREAS the Colonist and his family desire to settle on tillable land in the Matanuska Valley in the Territory of Alaska in order to obtain subsistence and gainful employment from the soil and coordinated enterprises, establish a home, and enjoy the benefits of the Rural Community now being formed there; and

WHEREAS the Corporation is a non-profit corporation and has been organized and established to assist worthy and well-qualified individuals and families to accomplish the above mentioned purposes and it desires to assist the Colonist and the members of his family in doing so;

THEREFORE BE IT AGREED, for and in consideration of the above premises and the mutual covenants herein contained, as follows:

1. TRANSPORTATION TO ALASKA

The Corporation will assume the obligation of the freight transportation of household and other effects up to two thousands (2,000) pounds of the Colonists and the above mentioned members of his family from the point of departure to Palmer Station in the Matanuska Valley, and advance and pay for the purchase of, and include in said freight and its transportation, such needed household furniture, small tools and home equipment as shall be agreed upon; same to be ultimately repaid by the Colonist a the same low cost and special Colonist rates as that charged to the Corporation.*

2. TEMPORARY CAMP

Upon arrival of the Colonist and his family at Palmer Station the Corporation will make available tents for their temporary shelter and habitation pending construction of their dwelling house and their moving on the land which they expect to make their permanent home.

3. LAND AND HOME IN THE RURAL COLONY

The Corporation will make available to the Colonist for a farm and home for himself and his family not less than forty (40) acres of land on terms of payment running over a period of thirty years.**

The Corporation will finance the Colonist in building his dwelling house and other permanent improvements on the land. The Colonist will repay for the same on an amortized plan over a period of thirty years.

Appendices

4. FARM MACHINERY AND EQUIPMENT

The Corporation will furnish the Colonist farm machinery, equipment, livestock and other supplies and furnishings on such use-charge, lease, rental or sale basis as may be agreed upon.

5. SUBSISTENCE

The Corporation will furnish subsistence to the Colonist and the above members of his family at actual cost from their arrival at Palmer Station until such time as the products which the Colonist and his family raise will enable him directly or by exchange or sale to furnish subsistence for himself and family.

6. COMMUNITY ACTIVITIES

The Corporation will build and equip such educational, cultural, recreational, health, work, and business centers in the community as the life of the community shall require, and make the same available to the Colonist and members of his family and other members of the community, and will furnish social and economic direction, supervisory and consultation services to the Colonist, members of his family and other members of the community on terms of mutual agreement and accord.

FULFULLMENT OF THIS AGREEMENT

The Colonists agrees that the relationship established by this contract between him and the Corporation is to assist him and the members of his family to become established in a new home on a self-sustaining and self-supporting basis, and that he will repay all loans made to him by the Corporation in connection with the provisions under the above numbered headings of this agreement or otherwise made to him by the Corporation, and pay for all material, supplies, equipment, furnishings, services, and personal, real, or mixed property referred to in the provisions under the above numbered headings of this agreement or otherwise furnished him by the Corporation, which are rented, leased, or sold to him by or through the Corporation, upon such terms as are agreed upon, and will enter into and perform all obligations and contracts necessary in order to do so; it being understood that interest rates on all obligations shall not be greater than three (3) per cent per annum from the time they are incurred and that payment of said interest shall not begin until the first day of September, 1938, and that payment of installments of the principal on all said obligations shall not begin until the first day of September, 1940 unless the Colonist elects to make such payment at an earlier date.

The Colonist further agrees that he and the members of his family will abide by all Corporation administrative directions and supervision in connection with control of crop production, processing, marketing, distribution, crop rotation, soil management, sanitation and other measures for the welfare of the community, and to cooperate with the Corporation, its representatives, and with the other colonists in building up a successful Rural Community.

It is mutually agreed by the parties hereto that this agreement is subject to any Federal, State or Territorial laws now existing or which may be hereafter enacted.

Appendices

-----

* The expense of travel of the colonists and the members of his family and the carriage of their baggafe from the point of departure to destination in Alaksa is to be attended to by the Emergency Relief Administration of the home state at no cost to the Colonists or members of his family and with no obligation of repayment.

** The Corporation is in a position to make available to the Colonists timbered land as low as five ($5.00) dollars and acre and other land at prices in proportion thereto depending upon the location and the extent to which the land has been cleared.

Appendix 3

Alaska Rural Rehabilitation Corporation Contract for Sale and Purchase of Realty

THIS CONTRACT made and entered into by and between ALASKA RURAL REHABILITATION CORPORATION, an Alaskan corporation with offices at Palmer, Alaska, in the Cooperative Community known as the Matanuska Valley Colony, hereinafter called the Corporation, and the undersigned purchaser, a member of said Cooperative Rural community, whose post office address is Palmer, Alaska, hereinafter called the Purchaser, who is joined in the execution of the same by the undersigned wife of the Purchaser, WITNESSETH:

WHEREAS, the Cooperative Rural Community known as the Matanuska Valley Colony, has been established at and near Palmer, Alaska, by the Corporation, by the aid of grants and funds of the Government of the United States made through the Federal Emergency Relief Administration in pursuance of public policy and for the public purpose of assisting in promoting and establishing colonization and rural rehabilitation in Alaska by enabling worthy and well qualified persons and their families to:

(a) occupy family sized tracts of land with adequate improvements and a home thereon in a Cooperative Rural Community and thereby obtain employment and gainful living in agricultural and allied activities and enjoy the benefits of said community under properly controlled conditions on a cooperative basis; and

(b) in due time to become the owners of said property on long time and low payment terms not ordinarily procurable through usual commercial channels, and thereby be placed in a position

Appendices

to enjoy the benefits referred to in sub-paragraph (a) herein permanently; and

WHEREAS, in the course of carrying out said public trust under the application of sub-paragraph (a) above the Purchaser and others have each been placed by the Corporation in occupancy of an improved family sized tract of land and in a home therein in said Cooperative Rural Community and the economic advantages referred to in sub-paragraph (a) are afforded the Purchaser and his wife and family and said other occupants of said tracts of land by membership in Matanuska Valley Farmers Cooperating Association, organized through the sponsorship of the corporation under the Cooperative Association laws of Alaska and in conformity with "The Agricultural Marketing Act" of the United States to enable the Purchaser and other members of said Community and their successors to cooperatively market their agricultural products raised upon the land occupied and farmed by them and to cooperatively purchase and otherwise acquire needed supplies, commodities and services through said association; and

WHEREAS, the Purchaser now wished the benefits provided for in sub-paragraph (b) above to be extended to him and the benefits enjoyed under sub-paragraph (a) above to be thereby made permanent and wishes to enter into a contract of purchase, in pursuance of said desires, with the Corporation and the said Corporation approves the entering into said contract;

NOW, THEREFORE, in consideration of the above premises and the mutual covenants and considerations herein mentioned, as well as and including the low purchase price, low installment payments extended over a long time period and other terms which are greatly to the advantage of the Purchaser as a Colonist, the Corporation and the Purchaser as parties hereto for themselves, their respective successors, assigns, heirs, executors, and/or administrators, AGREE AS FOLLOWS:

I. Property and Purchase Price. The Corporation in consideration of the principal sum of_____(_____) as the purchase price thereof with credits thereon as herein allowed and applied, with interest in the unpaid portions thereof to be paid as hereinafter provided, and of the agreements, covenants, conditions and reservations herein contained, hereby agrees to sell to the Purchaser and the Purchaser agrees to buy upon the terms and conditions herein provided the following described property situated in the Palmer Recording Precinct of the Third Judicial District of the Territory of Alaska:

[DESCRIPTION OF PROPERTY AND IMPROVEMENTS]

together with all the tenements, hereditaments and appurtenances thereto belonging or in any wise appertaining.

II. Use of Premises as Farm Home. The Purchaser shall personally and continuously occupy and use said above described premises as a farm and as a home for himself and the members of his family living thereon unless the Corporation shall otherwise consent thereto in writing and shall not establish any mercantile or similar business thereon; shall follow and observe such production, crop and distribution control and practices, including soil conservation and windbreak methods and conduct such livestock, poultry and dairy and other enterprises on said property as are in accordance with approved farm organization, management, practices and good husbandry, under the advice and direction of the Corporation and in compliance with his contract or contracts with Matanuska Valley Farmers Cooperating Association; shall at all times maintain said premises in good condition and repair and free from weeds, brush, washes or gulleys detrimental to efficient farming operations or the value of said premises for agricultural use; and shall not commit or permit any unlawful acts or waste or nuisances upon said premises or any part thereof.

Appendices

III. Care and Preservation of Property. The Purchaser shall not without the written consent of the Corporation: (a) demolish, alter or change the location of any of the buildings or structures or erect new ones on said premises; (b) in any way interfere with the easements or rights of way herein referred to or provided for; (c) lease or let any part of said premises; or (d) mortgage or encumber, other than to the Corporation or Matanuska Valley Farmers Cooperating Association, any crops or products of said premises or any machinery, equipment, tools, or livestock used by him in farming on said premises.

IV. *Terms of Payment and Security.* The Corporation hereby allows and applies to the payment of the principal sum named in Section I hereof as the purchase price of said property the following credits: _____
leaving the total amount which the Purchaser agrees to pay to complete the payment of said purchase price thereof the sum of _____ (_____) together with interest at the rate of three per cent (3%) per annum on the balance thereof from time to time remaining unpaid, the same being payable in the following installments and manner: the final payment of the last installment to be made on the ____ day of _____, 19__; provided that, if the date of any installment payment should fall on a legal holiday said installment payment, including the final payment, may be paid on the next business day following, and provided, also, that with the written consent of the Corporation the Purchaser shall have the privilege of making prepayments of any of said installments or portions thereof, and the privilege of making additional payments for the reduction or extinguishment of the principal debt and interest thereon on any installment date. Each and every installment not paid when due shall bear interest thereafter at the rate of three per cent (3%) per annum until paid which amount shall be added to said installment when paid. The Corporation expressly reserves a first lien and mortgage on the Purchaser's right, title and interest, now held

or hereafter acquired in the above described land and improvements thereon and on said land itself to secure the payment of any and all sums due hereunder and the fulfillment of any and all terms of this Contract and upon the crops grown upon said lands each year to secure the payment of any sums which may become due hereunder and be unpaid at any time the crop for such year is gathered or harvested and the Purchaser and his wife hereby mortgage the same accordingly to the Corporation.

The proceeds of any installment of payment made to the Corporation shall be applied as follows: First—to payment of any interest on unpaid balances of the principal sum due hereunder at the date of such installment; secondly—to the payment on account of the principal sum of this Contract; provided, however, that the Corporation may apply the proceeds of any installment or other payment made by the Purchaser hereunder to the payment of issuance, taxes, and/or assessments due or delinquent and payable by the Purchaser hereunder on or in connection with the above described premises before applying the same to the payment of interest or principal due and payable as herein above provided, in which case the amount of the installment remaining unpaid shall bear interest at the rate of three per cent (3%) per annum until paid the same as an unpaid installment as provided herein. The Purchaser also agrees that the Corporation may at its discretion apply to the payment of any obligations due under this Contract any moneys in its hands as agent or otherwise, of or due the Purchaser from any source whatsoever, except as otherwise agreed upon, but it is expressly mutually understood and agreed that the Corporation is not obligated to do so.

V. *Taxes and Assessments.* The Purchaser covenants and agrees to pay all taxes including Territorial and local taxes and any and all installments of special improvement taxes and/or assessments, including road, levee and drainage taxes due and payable from time to time on the above described premises, except as he

Appendices

is relieved from doing so by the homestead tax exemption laws or other laws of the Territory of Alaska. It is mutually understood and agreed that in the event the Purchaser fails to keep up and pay said taxes and assessments as herein above provided the Corporation may pay the same with interest at the rate ofthree per cent (3%) per annum from the date of such payment with interest and the amount or amounts thus paid by the Corporation may be added to the principal sum representing the purchase price hereunder, or the Corporation may pay the same and take the amount thereof out of the installment payment or payments made by the Purchaser with which to do so as provided for in Section IV hereof or for repayment to itself.

VI. *Insurance.* It is mutually understood and agreed that the Purchaser together with his family is in occupancy and in charge of the above described premises and that he is responsible for the safety and security thereof and must make good injuries or losses thereto. The Purchaser agrees upon written notice from the Corporation to take out fire, windstorm, hail and tornado insurance in the amount specified by the Corporation, payable to the Corporation in trust for the Purchaser, covering improvements on said premises in a firm, company or association approved by the Corporation and pay the premiums thereon. In the event of loss or damages recovered upon the policies, the proceeds, or so much thereof as may be necessary, shall be used to repair or restore the property damaged or destroyed. It is also mutually understood and agreed that in the event the Purchaser fails to keep up and pay his premiums on said insurance the Corporation may pay the same itself and the Purchaser shall be obligated to repay the same to the Corporation at the rate of three per cent (3%) per annum from the date of such payment until paid and the amount or amounts thus paid by the Corporation may be added to the principal sum representing the purchase price hereunder or the Corporation may pay the same and take the amount thereof out of the installment payment or payments made by the Purchaser with which to do so as provided for in Section IV hereof or for repayment to itself. It is mutually understood and agreed that until such time as the Purchaser is notified in writing to take out said insurance, the Corporation will take care of the protection of said property resulting from loss or damage by fire, which is not caused by the willful or negligent act of the Purchaser or his servants; provided, however, that this paragraph shall not be construed to make the Corporation liable for any household goods, or other personal property belonging to the Purchaser.

VII. *Public Trust Impressed upon the Land, and its Transfer.* It is mutually greed and recognized that the above described premises are subject to the fulfillment of the public trust and purpose for which the Federal grants of funds were made to purchase and acquire the same as part of the Cooperative Rural Community of Matanuska Valley Colony as set forth herein and more particularly in the "Whereas" provisions at the beginning of this Contract and that said trust and purpose include administrative regulations governing said Community and the occupancy and use of said premises according to and under the rules of the Matanuska Valley Farmers Cooperating Association, and that said trust and purpose thus subjecting said premises constitute a condition running with the land and improvements thereon, adhere to and follow said premises continuously into the hands of all holders or owners thereof forever whether they or any of them hold by grant, gift, sale, devise, inheritance or otherwise. It is further mutually agreed and recognized that said premises or any right, title, interest or privilege therein is subject to purchase by the Corporation at its option at any time the same passes by death, operation of law, or otherwise to any person or persons (natural, corporate or governmental), incapable, unable, unwilling, or unfit to comply with and carry out the requirements and provisions of this Section of this Contract in reference to said trust and purpose or who refuse to do so as determined by the General Manager of the Corporation. The Purchaser agrees that

Appendices

he will not grant, make a gift of, bargain, sell, devise or otherwise convey, assign, or dispose of said premises or his interest therein or any part thereof or in this Contract or any part thereof or attempt to do so to anyone other than the Corporation who is not desirous and capable of complying with the requirements and conditions of this Section of this Contract and who does not sign a written agreement so to do and to become a bona fide member of said Community and of Matanuska Valley Farmers Cooperating Association on a farm furnished by the Corporation and who is not approved in writing as acceptable as such by the General Manager of the Corporation. Provided, however, that the Purchaser may grant, make a gift, or bargain, sell, devise or otherwise convey, transfer, assign or dispose of said premises or his interest therein or any part thereof or in this Contract or any part thereof to the Corporation and the Corporation may buy the same at any time the Corporation and the Purchaser shall agree to do so as provided in this Contract as it is mutually agreed that the Corporation shall have the first right and option to buy the same and if the Purchaser desires to sell and receives a bona fide offer from a permitted prospective buyer hereunder the offer together with the terms thereof must be reduced to writing, signed by the offerer and properly acknowledged before a Notary Public and shown to the General Manager of the Corporation and it it is an enforceable offer at a higher price and on better terms than the Corporation agrees to pay and give, then the General Manager, in behalf of the Corporation, shall, at the request of the Purchaser herein, relinquish the Corporation's first right to buy by endorsement accordingly on the said written offer signed by him and properly notarized and then the Purchaser herein may sell to the said offerer on the terms of said offer only, all legal documents of sale thereunder being subject to inspection of the General Manager of the Corporation and attempted sales other than under the terms of said offer being null and void, subject to prevention by injunction and constituting a fundamental breach of this Contract.

It is mutually agreed and recognized that this Section of this Contract is in pursuance of public policy and the welfare of the Rural Cooperative Community of Matanuska Valley Colony and the members thereof and to assure the fulfillment of the purposes of said federal grants of funds as provided herein and for the protection of the continued use and occupancy of said premises in accordance therewith. It is also mutually agreed and recognized that this Section of this Contract is a principal part of the consideration thereof and of the Corporation entering into the same and its agreement to part with the title to the said premises for the low purchase price and is enforceable by specific performance, injunction and mandamus and that any attempt to sell or transfer said premises or interest therein or in this Contract in violation of the terms hereof shall be and is null and void.

VIII. *Corporation's Right to Buy.* In the event the Corporation shall purchase or exercise its right and option to purchase the Purchaser's interest in this Contract and the above described premises pursuant to Section VII hereof the Corporation shall pay the Purchaser and the Purchaser agrees to accept in full payment thereof an amount computed as follows, to-wit:

(a) There shall be added to the total amount of principal paid by the Purchaser prior to the date of the exercise of the above mentioned option (1) the then fair value of any crops planted by the Purchaser and then growing on said premises and (2) the then fair value of any improvements made by the Purchaser on said premises with the consent of the Corporation.

(b) From the amount calculated in (a) above there shall be deducted (1) any interest, insurance, taxes or assessments then accrued and unpaid and (2) any depreciation in the value of said premises caused by the use thereof, of by failure of the Purchaser to maintain said premises in good condition and repair, (3) any debts due from the Purchaser to the Corporation and any obligations whatsoever.

Appendices

In case the Corporation and the Purchaser shall not agree on the fair value of any such crops or improvements or the amount of any such depreciation the same shall be determined by the majority of three appraisers, one of whom shall be appointed by the General Manager of the Corporation, one by the Purchaser, and the third by the other two appraisers so appointed. The method of computation of purchase price provided in this Section shall be applied in all cases of purchases made by the Corporation from any person or persons under the authority of Section VII hereof, and when a portion of a right, title, interest or privilege in said premises is purchased, said method of computation of purchase price shall be applied proportionately according to the portion purchased.

IX. *Purchaser's Relationship with the Community, the Cooperative Association and the Corporation.* The Purchaser now agrees that he has entered into this Contract with the purpose in view of establishing permanent home for himself and family on the premises described herein in the Cooperative Rural Community of Matanuska Valley Colony and not for resale or speculation but to enjoy the benefits and to help fulfill the public trust and public policy and purpose for which Federal grants of funds have been made and have been and are being expended by the Corporation as more particularly set forth in the "Whereas" provisions at the beginning of this Contract as a part hereof, and that he and the members of his family will abide by all administrative directions and supervisions of the Corporation, its duly constituted agents and representatives properly authorized and keep and perform their obligations to and contract with the Corporation.

The Purchaser further agrees to accept and maintain membership in Matanuska Valley Farmers Cooperating Association, on the terms provided by said association, and to keep and perform his contract and membership obligations therewith, and that he and the members of his family will abide by and comply with its rules and regulations and cooperate with the other members of said Community and said Association and with the Corporation in all activities within the Community and in connection therewith for the success thereof.

The Purchaser further agrees that the Corporation and its duly authorized agents shall have the right of ingress and egress at all reasonable times, over, across, and upon said premises herein described for the purpose of inspection, supervision, maintenance and carrying out said rules and regulations and the public trust and public policy and purpose for which the Cooperative Rural Community of Matanuska Valley Colony has been established, and in order to protect, help improve, and to conserve said premises, its interest therein and in this Contract, and that the Corporation also has easement and right of way in, on, and over said premises, for the construction, operation and maintenance of light, water, power, telephone and other utilities commonly known as public utilities (whether publicly or privately owned) which easement and right of way the Corporation especially reserves.

X. *Violation of Contract.* In the event that the Purchaser violates any of the terms or conditions of this Contract, or is adjudicated a bankrupt or an insolvent, the Corporation shall have the right to terminate all of the rights of the Purchaser hereunder, by giving written notice addressed to the Purchaser that his rights hereunder shall cease and determine thirty (30) days subsequent to the date of such notice; and upon the expiration of thirty (30) days specified in said notice, the rights of the Purchaser hereunder shall cease and determine absolutely, except as herinafter otherwise expressly provided.

XI. *Termination Rights of Purchaser.* In the event the rights of the Purchaser under this Contract shall be terminated in accordance with the provisions of Section X, the Corporation shall, without waving any of its other rights hereunder, cancel this Contract and pay to the Purchaser within sixty (60 days) after

Appendices

the Purchaser shall have vacated possession in accordance with the provisions of Section XII, a sum equal to the amount which would be payable by the Corporation if it exercised its right of purchase provided for in Section VII above, computed as of the date the Purchaser vacates possession, less any expenses incurred by the Corporation, including all Court costs, Attorneys' fees, or other expenses necessitated by the Purchaser's breach or failure to comply with any of the provisions of this Contract, and/or in the Corporation regaining possession of the above described premises under Section XII of this Contract, by Court action or otherwise, and/or for the enforcement of any right of the Corporation hereunder.

XII. *Corporation's Right of Repossession.* In the event of the termination of the rights of the Purchaser under this Contract or the Corporation exercising its right to purchase or its acceptance of the offer of the Purchaser to sell his interest in this Contract and in the above described premises as provided in this Contract, the Purchaser shall, within thirty (30) days after receipt of notice from the Corporation of the same, vacate and surrender possession of said property. In the event the Purchaser fails to vacate and surrender possession of said property within such time, he shall remain in possession thereof only as a tenant at the sufferance of the Corporation and the Corporation shall have the right and power immediately or at any time thereafter to re-enter upon and take possession of said property and remove therefrom the Purchaser or anyone whether claiming by, through or under him or otherwise and their property and effects without process of law and without being guilty of any manner of trespass either in law or in equity. Provided, however, that it is mutually understood and agreed by both parties to this Contract that if the above described land is public land occupied by the Purchaser by his lawful settlement and entry thereon and in fulfillment of other governmental requirements and not yet proved upon by the Purchaser and he has not received a patent and title to said land from the Government at the time of any violations of this Contract by the Purchaser should take place, then the Corporation, at its election shall exercise its right of repossession by removing the above described buildings and improvements from said land as personality and same shall in no ways be a trespass, and may foreclose the Purchaser's interest, if any, in the same as provided herein.

XIII. *Final Deed of Conveyance.* Upon the payment in full by the Purchaser of all sums due the Corporation under the provisions of this Contract and in the event that this Contract shall not have been heretofore terminated under the terms thereof by the Corporation, the Corporation shall execute and deliver to the Purchaser a deed of conveyance to the premises herein described, which shall recite the conditions of Section VII hereof continuously running with the land and improvements thereof, together with provisions for the reversion of title to said premises to the Corporation in case of breach thereof, as a part and condition of said deed of conveyance and warrant and convey a good and merchantable title to the premises therein described free of encumbrances to the Purchaser, his heirs and assigns forever.

XIV. *Effect of this Contract.* It is mutually agreed and understood that this Contract is binding upon and endures to the successors, assigns, heirs, executors, administrators and legal representatives of the respective parties hereto and that there are no oral or other conditions, promises, covenants, representations or inducements in addition to or at variance with any terms hereof and that this Contract expresses the voluntary and clear understanding of the parties hereto fully and completely, and that it goes into effect on the _____ day of _____.

Appendix 4
Project Costs

A. Estimated Cost for Project*

Buildings:

200 houses, including wells & fencing	$232,000
200 barns	40,000
Implement barn	1,080
Garage and repair shop	8,918
Creamery	15,750
Cannery	8,160
Poultry houses for community hatchery	6,000
Manager's house	4,360
Teacher's home, equipped	34,200
2 Grain-mill-work center buildings	2,300
Warehouse 40' x 120' x 10'	4,284
Duplex teachers' house	3,260
Trading post	10,000
School and community house	150,000
Subtotal	**$ 522,512**

Equipment:

Garage and repair shop	$5,500
Implements for rental and construction	103,130
Work center-grain mills	1,550
Creamery	6,200
Cannery	7,200
Poultry houses	2,000
Central heating plant	18,000
Subtotal	**$143,580**

Supplies:

Livestock --
400 cows	$54,000
5 bulls	1,000
3,000 hens	3,000
100 horses	17,500
50 sets harnesses	1,750

*Cost estimate appears in memo to Minnesota ERA dated March 13, 1935 from FERA.

50 Army wagons	$2,750
Furniture and household goods	57,000
Seed	6,000
Feed for horses and 100 cows—1st year	17,815
Farm tools and equipment	15,000
Subtotal	**$155,815**

Land:

Cost of accumulation	$500
Surveying and staking	5,000
2200 A	23,000
Deeds and filing fees	1,500
Subtotal	**$30,000**
Subtotal, all facilities	**$851,907**
Transportation	60,000
Operating capital	60,000
Supervision during construction	10,000
Total estimate cost	**$981,907**

Appendices

B. Estimated cost per Family*

40 acre farm	$300.00
Community plan covering clearance & cultivation	210.00
House and well	1,600.00
Barn	200.00
Fences	60.00
Livestock and Poultry	345.00
Furniture and household articles	285.00
Farm equipment and tools	75.00
Seed and feed	130.00
Railroad and boat fare per family	300.00
Surveying charge	7.50
Total	**$3,512.50**

* Cost estimate appears in memo to Minnesota ERA dated March 13, 1935 from FERA.

C. Actual Cost for Project Grants from FERA*

Grants to: ARRC	$3,930,718.68
California ERA	716,907.36
Montana ERA for cows and horses	19,782.67
Wyoming ERA for horses	14,447.84
Subtotal - grants to ARRC	**$4,681,856.55**
Michigan ERA	26,823.51
Minnesota ERA	30,347.66
Wisconsin ERA	20,195.49
Alaska Road Commission	648,466.00
Total granted for colonization	**$5,407,689.21**

* Stone, Kirk H., The Matanuska Valley Colony, 1950, page 41, table 9.

Graphic Image Credits

Page iii. Miller family photo—1940, Margaret Miller Photo Collection, Palmer Museum of History and Art

Page 6. ARRC staff—1935. Ibid.

Page 8. Matanuska Experiment Station, Ibid.

Page 9. Another view of Matanuska Experiment station, Ibid.

Page 16. Miller Family photo—1935, Ibid.

Page 24. USAT St. Mihiel, Ibid.

Page 26. City of Seward, Ibid.

Page 27. Aerial view of Anchorage, Ibid.

Page 29. Palmer post office & store, Mary Nan Gamble photographs, Willis T. Geisman, P270-591 Alaska State Library

Page 30. Aerial photo of Palmer—1935, Margaret Miller Photo Collection, Palmer Museum of History and Art

Page 32. Camp 7, Mary Nan Gamble photographs, Willis T. Geisman, P270-161 Alaska State Library

Page 36. Wash day at Camp 7, Margaret Miller Photo Collection, Palmer Museum of History and Art

Page 39. Palmer and old ARRC warehouse—1935, Ibid.

Page 44. Pioneer Peak, Mary Nan Gamble photographs, Willis T. Geisman, P270-667, Alaska State Library

Page 45. Talkeetna Mountains, Ibid., P270-532

Page 49. Display log cabin, Ibid. P270-250.

Page 53. Map of Alaska from *Milwaukie Journal*, Margaret Miller Papers, in possession of family

Page 55. Rex Beach headline from *Saturday Evening Post*, Ibid.

Page 59. Senate opens Quiz headline from *Milwaukie Journal*, Ibid.

Page 63. Congressman Zioncheck, Mary Nan Gamble photographs, Willis T. Geisman, P270-848 Alaska State Library

Page 65. Samuel Fuller and entourage, Ibid., P270-837

Page 69. Bert Bingle and fishing crew, Margaret Miller Photo Collection, Palmer Museum of History and Art

Page 70. Margaret and other colonist wives, Margaret Miller Photo Collection, Palmer Museum of History and Art

Graphic Image Credits

Page 71. Colonist picnic at Wasilla Lake. Ibid.
Page 72. Haircut day at Camp 7, Ibid.
Page 76. Sawmill at Camp 5, Ibid.
Page 84. Neil and Margaret digging cellar, Ibid.
Page 88. HarrietMalstrom, Mary Nan Gamble photographs, Willis T. Geisman, P270-624, Alaska State Library
Page 92. School construction, Almer J. Peterson papers & photographs, Univ. of AK Anchorage. Consortium Library. Archives & Special Collections
Page 93. The Miller house in 1935, Margaret Miller Photo Collection, Palmer Museum of History and Art
Page 99. Adam and Fannie Werner homestead, Margaret Miller Photo Collection, Palmer Museum of History and Art
Page 101. Bachelor settler tract, Ibid.
Page 105. Mary Rossiter & twins, Mary Nan Gamble photographs, Willis T. Geisman, P270-659, Alaska State Library
Page 108. Neil's bus, Margaret Miller Photo Collection, Palmer Museum of History and Art
Page 111. Colonel Leroy Hunt, Mary Nan Gamble photographs, Willis T. Geisman, P270-829, Alaska State Library
Page 197. The Miller farm in winter, Margaret Miller Photo Collection, Palmer Museum of History and Art
Page 123. School bus on muddy road, Ibid.
Page 125. Well-drilling rig, Mary Nan Gamble photographs, Willis T. Geisman, P270-929, Alaska State Library
Page 131. Palmer community center, Margaret Miller Photo Collection, Palmer Museum of History and Art
Page 142. View of Anchorage in winter, Ibid.
Page 146. Palmer skating rink, Ibid.
Page 149. Market Street in winter, Ruby I. Cashen papers, Univ. of AK Anchorage. Consortium Library. Archives & Special Collections
Page 156. Fourth Avenue in Anchorage, Margaret Miller Photo Collection, Palmer Museum of History and Art
Page 158. The Miller's first cow, Arabella, Margaret Miller Photo Collection, Palmer Museum of History and Art

Page 163. Warehouse fire, Almer J. Peterson papers and photographs, Univ. of AK Anchorage. Consortium Library. Archives & Special Collections

Page 167. Corporation tractor clearing stumps, Margaret Miller Photo Collection, Palmer Museum of History and Art

Page 169. View of Palmer from the west, Margaret Miller Photo Collection, Palmer Museum of History and Art

Page 174. ARRC administrative building, Ibid.

Page 175. ARRC hospital, Ibid.

Page 176. ARRC cannery and creamery, Ibid.

Page 177. ARRC commissary/trading post, Ibid.

Page 178. Market Street in spring, Ibid.

Page 181. Harvesting hay on the hay flats, Mary Nan Gamble photographs, Willis T. Geisman, P270-819, Alaska State Library

Page 182. Baseball game in Palmer, C. Earl Albrecht papers, Univ. of AK Anchorage. Consortium Library. Archives & Special Collections

Page 184. Stock barns at fairgrounds, Margaret Miller Photo Collection, Palmer Museum of History and Art

Page 185. Anchorage to Palmer highway, Margaret Miller Photo Collection, Palmer Museum of History and Art

Page 185. Highway bridge over Knik River, Ibid.

Page 190. UPC, Margaret Miller Photo Collection, Palmer Museum of History and Art Stock barns – 191

Page 198. Miller truck, Margaret Miller Photo Collection, Palmer Museum of History and Art

Page 201. Neil and Tim doing chores, Ibid.

Page 207. The Puhl farm, Ibid.

Page 208. Neil in the fields, Ibid.

Front cover. Miller house in 1935, Margaret Miller Photo Collection, Palmer Museum of History and Art

Back cover. Miller family pausing during work on their house to talk with reporter, Mary Nan Gamble photographs, Willis T. Geisman, P270-612, Alaska State Library

Notes

Prologue

1. USDOI, Kirk Stone, 1950, pp 41.
2. O.W. Miller, 1975, pp 62-63.
3. Ibid., pp 42-43.
4. U.S. Census, 1910, 1920.
5. Helen Hegener, 2017, p 230; also U.S. Census, 1920.
6. U.S. Census, 1930.
7. Don Irwin, 1968, p 2; also U.S. Geological Survey, Richard G. Ray, 1904, pp 35-36; Also U.S.G.S., Sidney Paige and Adolph Knopf. 1907, pp 116-118; Also U.S.G.S., Stephen R. Capps, 1917, p 117.
8. Helen Hegener, 2017, p 130; also Clarence Hulley, 1953, p 323
9. Alaska Agricultural Experiment Station, Hugh A. Johnson and Keith L. Stanton, 1955.
10. USDOI, Kirk Stone, 1950, pp 22-24.
11 Homesteading refers to the federal program authorized under the 1862 Homestead Act that allowed individuals to enter unoccupied federal lands that had been opened to homesteading, continuously live on those lands for a pre-determined length of time, and improve those lands for use as farms. If individuals fulfilled the requirements of the act (which included paying entry and final patent fees) they could receive title to the land without purchase. Homesteading on federal lands in Alaska ended on October 20, 1986.
12. O.W. Miller, 1975, p 74: also USDOI, Kirk Stone, 1950, Table 3
13. Ibid.,, 1975, p 75; Also Hess Papers, ARRC articles of incorporation.
14. O.W. Miller, 1975, p 55.
15. Hess Papers, Memo to Minnesota ERA from FERA officials in Washington, D.C. attached as note to June 18, 1937 letter from H.M. Colvin to Colonel Lawrence.

Westbrook, Assistant Administrator, Works Progress Administration; also O.W. Miller, 1975, pp 69-72, also Don Irwin, 1968, pp 62-64.
16. O.W. Miller, 1975, p 72.
17. Ibid., pp 74-75, also Don Irwin, 1968, pp 53-54, also ARRC files.
18. Figures on number of transients disagree slightly. USDOI, Kirk Stone, 1950, p 50, says 420 transients. Marvin Halldorsen, 1936, p 95, says 425 transients.
19. Fairbanks Daily News Miner, April 18, 1935.
20. Evangeline Atwood, 1966, p 181.
21. Ibid., 1966, p 182.

Chapter 1

1. Evangeline Atwood, 1966, pp 46-47.

Chapter 2

1. O.W. Miller, 1975, pp 69-72.
2. Don Irwin, 1968, p 58, also Evangeline Atwood, 1966, p. 62.

Chapter 3

1. *Anchorage Times*, July 19, 1935.
2. O.W. Miller, 1975, p 101; also USDOI, Kirk Stone, 1950, p 13.
3. O.W. Miller, 1975, p 75; also USDOI, Kirk Stone, 1950, p 41.
4. Don Irwin, 1968, p 60.
5. *Fairbanks Daily News Miner*, May 8, 1935.
6. Don Irwin, 1968, p 66-67.
7. Ibid., p 67; also Miller Papers, colony council minutes for June 18, 1935.
8. Don Irwin, 1968, pp 59-60; also USDOI, Kirk Stone, 1950,

Notes

p 75; also *Fairbanks Daily News Miner*, June 17, 1935.
9. USDOI, Kirk Stone, 1950, p 16; also Miller Papers, colony council minutes for June 14, 1935.
10. USDOI, Kirk Stone, 1950, p 2.
11. Don Irwin, 1968, p 57; also University of Alaska, Watson, Branton, Newman, 1971, p 23; also USDOI., Dale, 1956, p 6.
12. Don Irwin, 1968, p 57.
13. O.W. Miller, 1975, p 79; also USDOI., Kirk Stone, 1950, p 52; also Marvin Halldorsen, 1936, p 95.
14. Don Irwin, 1968, p 100.
15. USDOI, Kirk Stone, 1950, pp 55-56; also Miller Papers, colony council minutes for June 4, 1935.
16. Evangeline Atwood, 1966, p 130.
17. Ibid., p 130; also Bert Bingle, 1938, pp 10-13.

Chapter 4

1. O.W. Miller, 1975, pp 144-145.
2. Ibid., p 144.
3. Miller Papers, *Saturday Evening Post* article in scrapbook.
4. *Fairbanks Daily News Miner*, June 14, 1935.
5. O.W. Miller, 1975, p 84.
6. USDOI, Kirk Stone, 1950, pp 35-36.
7. Miller Papers, St. *Paul Pioneer Press* article in scrapbook.
8. Ibid, *St. Paul Pioneer Press* article in scrapbook.
9. Ibid. *River Falls Journal* article in scrapbook.
10. Ibid.; also *Fairbanks Daily News Miner*, June 12, 1935.
11. *Anchorage Times*, July 11, 1935.
12. *Fairbanks Daily News Miner*, May 27, 1935.
13. Williams Papers, June 24 letter from Harry Hopkins to President of U.S. Senate.
14. Miller Papers, article in scrapbook.
15. Williams Papers, personal letter dated July 2, 1935.

16. Williams Papers, undated personal letter; also Don Irwin, 1968, p 74.
17. Ibid., report from Samuel Fuller to Colonel Westbrook dated August 2, 1935.
18. Ibid., report from Samuel Fuller to Colonel Westbrook, August 5, 1935.
19. Miller Papers, *River Falls Journal* article in scrapbook.

Chapter 5

1. *Anchorage Times,* June 24, 1935.
2. Williams Papers, personal letter, May 20, 1935.
3. O.W. Miller, 1975, p 92.
4. Don Irwin, 1968, p 67.
5. Williams Papers, report from Samuel Fuller to Colonel Westbrook, August 5, 1935.
6. Ibid., personal letter, May 20, 1935.
7. Ibid., report from Samuel Fuller to Colonel Westbrook, August 5, 1935.
8. Ibid., personal letter dated July 17, 1935.
9. Ibid., report from Samuel Fuller to Colonel Westbrook, August 5, 1935; also O.W. Miller, 1975, p 86.
10. *Matansuka Valley Pioneer*, September 26, 1935, p 4; April 9, 1936, p 2.
11. *Alaska Weekly*, June 6, 1935; also O.W. Miller, 1975, pp 92-97.
12. O.W. Miller, 1975, p 95.
13. *Matanuska Valley Pioneer,* August 22, 1935, p 1; August 29, 1935 p 3.

Chapter 6

1. *Matanuska Valley Pioneer*, September 5, 1935. p 1.
2. Ibid., October 3, 1935, p 1.
3. Ibid,. September 3, 1935, p 5; October 3, 1935, p 5; October 17, 1935 p 2; November 14, 1935, p 4.

Notes

4. Don Irwin, 1968, p 78; also *Matanuska Valley Pioneer*, September 19, p 1.
5. *Matanuska Valley Pioneer*, October 17, 1935. p 4; October 24, p 2
6. Ibid, September 12, 1935, p 4.
7. Hess papers, telegram from Colonel Westbrook to Sheely, November 2, 1935; also minutes of ARRC board meeting, May 16, 1936; also minutes of ARRC board meeting, April 23-24, 1937; also *Matanuska Valley Pioneer*, May 21, 1936, p 2; July 30, 1936, p 1; December 16, 1936, p 1.
8. *Anchorage Times*, November 11, 1935; also USDOI, Kirk Stone, 1950, p 72.
9. Don Irwin, 1968, pp. 85-86.

Chapter 7

1. *Anchorage Times*, October 11, 1935; also University of Alaska, Watson, Branton, Newman, 1971, p 23; also U.S.D.O.C., Dale, 1956, p 6.
2. *Matanuska Valley Pioneer*, October 3 , 1935, p 5.
3. Ibid., November 14, 1935, p 4.
4. Ibid, October 31, November 28, 1935, p 1.
5. Ibid, Octtober 17, 1935, p 1.
6. Ibid, October 24, 1935, p 1.
7. Williams Papers, unpublished article by Williams titled "Matanuska Valley, the Latest Frontier.".
8. O.W. Miller, 1975, p 91; also Hess Papers, minutes of ARRC board of directors meeting October 29, 30, 1935.
9. Hess Papers, telegram from Ross Sheely to Colonel Westbrook November 11, 1935; also USDOI, Kirk Stone, 1950, pp 35-36.
10. *Matanuska Valley Pioneer*, October 17, 1935, p 2; October 24, 1935, p 4; November 28, p 2.
11. Williams Papers, undated memo from Leo Jacobs to Williams.
12. Marvin Halldorsen, 1936, p 111.

Chapter 8

1. *Matanuska Valley Pioneer*, Dec. 5, 1935, p 1.
2. Ibid, December 12, 1935, p 1.
3. Ibid, November 28, 1935. p 3.
4. Ibid, December 12, 1935, p 4.
5. University of Alaska, Watson, Branton, Newman, 1971, p 23; also USDOC., Dale, 1956, pp 6.
6. *Matanuska Valley Pioneer*, Dec 5, 1935, p 2.
7. Ibid, January 2, 1936, p 1; also Seward Gateway, December 24, 1935.
8. *Seward Gateway*, December 21, 1935.
9. *Matanuska Valley Pioneer*, January 2, 1935, p 1.
10. Don Irwin, 1968, p 83.

Chapter 9

1. Hess Papers, telegram from Ross Sheely to Colonel Westbrook dated January 4, 1936.
2. *Matanuska Valley Pioneer*, January 30, 1936. p 1.
3. *Fairbanks Daily News Miner*, February 2, 1936.
4. *Matanuska Valley Pioneer*, January 2, 1936, p 4.
5. *Fairbanks Daily News Miner*, January 17, 1936.
6. Hess Papers, letter from Ross Sheely to Luther Hess dated January 15, 1936.
7. Ibid., letter from Ross Sheely to Luther Hess dated January 16, 1936.
8. *Matanuska Valley Pioneer*, Jan. 16, p 4.
9. United States Geological Survey, Frank W. Trainer, 1963
10. *Matanuska Valley Pioneer*, January 9, 1936, p 2; also *Fairbanks Daily News Miner*, January 11, 1936.
11. University of Alaska, Watson, Branton, Newman, 1971, p 23; also USDOC, Dale, 1956, p 6.

Notes

Chapter 10

1. *Matanuska Valley Pioneer*, February 8, 1936, p 1.
2. *Fairbanks Daily News Miner*, February 10, 1936.
3. *Matanuska Valley Pioneer*, Dec. 19, p 4; Jan. 2, p 7; Feb 13., p 3.
4. *Matanuska Valley Pioneer*, February 20, 1936, p1.
5. Univ. of Alaska, Watson, Branton, Newman, 1971, p 23; also USDOC, Dale, 1956, p 6.
6. USDOI, Kirk Stone, 1950, pp 35-36.
7 *Matanuska Valley Pioneer*, February 6, 1936. p 2
8. Ibid, February 13, 1936. p 3.
9. Ibid., February 27, 1935. p 2. This article states that two additional men drew for tracts at the end of February, bringing the total number of families joining the project in February to 11. However, project records indicate the number of families joining in February was 10.
10. Ibid., February 13, p 2; February 20, 1936, p 2, 5.
11. Ibid., February 20, 1936, p 3.
12. Ibid., February 27, 1936 p 1.
13. USDOI, Kirk Stone, 1950, p 43.
14. *Seattle Post Intelligencer*, July 8, 1936.

Chapter 11

1. *Fairbanks Daily News Miner*, March 3, 5, 6, 1936.
2. *Matanuska Valley Pioneer*, March 6, 1936, p 2; March 12, p 1.
3. Ibid., April 25, 1936, p 3; also Don Irwin, 1968, p 103.
4. Ibid., March 19, 1936, p 7.
5. Theodore Feldman, 1941, 24-25.
6. *Matanuska Valley Pioneer*, March 6, 1936, p 1.
7. Univ. of Alaska, Watson, Branton, Newman, 1971, p 23; also USDOC, Dale, 1956, p 6.

8. *Matanuska Valley Pioneer*, March 12, 1936, p 5.
9. Ibid., February 27, 1936, p 3.
10. Ibid., March 6, p 5; March 12, 1936, p 1, 3; also Hess Papers, telegrams from Ross Sheely to Colonel Westbrook, April 4, April 25, 1936.
11. *Matanuska Valley Pioneer*, December 12, 1935, p 3.
12. Ibid., April 25, 1936, p 1.
13. USDOI, Kirk Stone, 1950, pp 35-36.
14. *Matanuska Valley Pioneer*, March 19, p 7; April 9, p 2; April 30, 1936, p 2; also Hess Papers, telegrams from Ross Sheely to Colonel Westbrook, April 4, April 25, May 2, 1936.

Chapter 12

1. O.W. Miller, 1975, p 122; also USDOI, Kirk Stone, 1950, p 72.
2. *Matanuska Valley Pioneer*, June 25, 1936, p 6.
3. USDOI, Kirk Stone, 1950, pp 65-66.
4. *Matanuska Valley Pioneer*, June 4, 1936. p 1.
5. Ibid., May 28, 1936, p 1.
6. Hess Papers, minutes from ARRC board of directors meeting, May 16, 1936.
7. *Matanuska Valley Pioneer*, April 16, 1936. p 5.
8. Hess Papers, minutes from ARRC board of directors meeting May 16, 1936.
9. O.W. Miller, 1975, p 113.
10. *Matanuska Valley Pioneer*, May 21, 1936. p 1.

Chatper 13

1. USDOI, Kirk Stone, 1950, pp 72–75.
2. *Matanuska Valley Pioneer*, May 28, p 1.
3. Ibid., June 25, p 4.
4. Ibid, August 6, 1936. p 6.
5. Ibid., June 11, 1936, p 1, 2; June 25, 1936. p 1; July 9, 1936, p 3.

Notes

6. Ibid, August 6, 1936. p 3.
7. Don Irwin, 1968, pp 102-103.
8. USDOI, Kirk Stone, 1950, p 37.
9. Ibid., pp 35-36.
10. O.W. Miller, 1975, p 128.
11. Ibid., p 137.
12. *Matanuska Valley Pioneer*, Nov. 19, 1936, p 2.
13. Ibid., Dec. 19, 1936, p 1.
14. O.W. Miller, 1975, p 123–124.
15. Ibid., 1975, p 125.
16. Ibid., 1975, pp 149-151.
17. Hess Papers, minutes from ARRC board of directors meeting, August 17, 1937.
18. USDOI, Kirk Stone, 1950, p 74.
19. Ibid., Kirk Stone, 1950, pp 35-36.
20. ARRC files, 1937 FERA report titled, "Matanuska Valley Resettlement Project."

Chapter 14

1. *Matanuska Valley Pioneer*, Nov. 12. 1935, p 1; Nov. 19, 1936, p1; also *Valley Settler*: Jan. 16, 1937, p 1; Jan. 6, 1938, p 8; Dec. 30, 1938, p 4.
2. Hess Papers, minutes from ARRC board of Directors meeting August 17, 1938.
3. *Valley Settler*, Sept. 16, 1938, p 1.
4. Ibid., Sept 30, 1938 p 2; Oct. 14, 1938, p 1.
5. Ibid., Oct. 14, 1938, p 1.
6. Ibid., Nov. 4, 1938, p1; Nov. 18, 1938, p 2.
7. USDOI, Kirk Stone, 1950, pp 35-36.
8. *Valley Settler*, Jan. 27, 1939, p 1; Feb. 3, 1939, p 2.
9. Ibid., Feb. 3, 1939, p2, Feb. 10, 1939, p 1.
10. O.W. Miller, 1975, p 154.
11. USDOI, Kirk Stone, 1950, p 73.
12. *Valley Settler*, Feb. 17, 1939, p 2; Feb, 24, 1939, p 1, p 5.

13. Hess Papers, minutes from ARRC board of Directors meeting August 21, 1939.
14. Ibid., minutes from ARRC board of Directors meeting, August 14, 1940.
15. *Valley Settler*, Jan. 12, 1940, p 3.
16. O.W. Miller, 1975, pp 132-136.
17. USDOI, Kirk Stone, 1950, pp 73, 75.
18. Hess Papers, minutes from ARRC board of Directors meeting, August 14, 1940, ARRC report July, 1942.
19. USDOI, Kirk Stone, 1950, p 82.
20. Hess Papers, minutes from ARRC board of Directors meeting, August, 1941.
21. O.W. Miller, 1975, p 141; *Valley Settler*, Oct.17, 1941, p 2.

Bibliography

Alaska Development Board. *The Matanuska Valley*, Circular No. 3. Juneau, Alaska: 1946.

Alaska Rural Rehabilitation Corporation. *Twenty Years of Progress in the Matanuska Valley, Alaska*. Palmer, Alaska: 1955.

Alaska Rural Rehabilitation Corporation office, Palmer, Alaska. Business files containing financial information on each tract in project.

Alaska Rural Rehabilitation Corporation records, in Archives and Special Collections, Consortium Library, University of Alaska, Anchorage. Records include files kept from 1935 to about 1945 on each colonist and other aspects of project.

Allman, Jack. *Matanuska Valley Pioneer*. 1935-1936.

Atwood, Evangeline. *We Shall be Remembered*. Anchorage, Alaska: Alaska Methodist University Press, 1966.

Bingle, Bert. *The First Three Years*. Palmer, Alaska: self-published, 1938.

Feldman, Theodore F. *The Federal Colonization Project in the Matanuska Valley, Alaska*. Masters thesis, University of Washington: 1941.

Fox, James H. *The First Summer: Photographs of the Matanuska Colony of 1935*. Alaska Rural Rehabilitation Corporation, 1980

Halldorsen, Marvin A. *The Matanuska Valley Colonization Project*. Masters thesis, University of Colorado: 1936.

Hegener, Helen, *The Alaska Railroad: 1902-1923*. Northern Light Media, 2017.

Hegener, Helen, *The 1935 Matanuska Colony Project: The Remarkable History of a New Deal Experiment in Alaska*. Northern Light Media, 2014.

Hess, Luther C., Papers. University of Alaska Archives, Fairbanks, Alaska. Six folders of material relating to the Matanuska Colony. Hess was vice-president of the ARRC board of Directors.

Hulley, Clarence C. *Alaska, Past and Present*. Portland, Oregon: Binford and Mort, 1953.

Irwin, Don L. *The Colorful Matanuska Valley*. Palmer, Alaska: self-published, 1968.

Jacobs, Leo. *Matanuska Progress*. Anchorage, Alaska: Anchorage Chamber of Commerce, no date (probably published in 1938).

Matanuska Valley Farmers Cooperating Association. *The Valley Settler*. 1937-195?

Miller, Margaret. Papers. In possession of family, Fairbanks, Alaska. Correspondence, journals, council minutes, newspaper clippings, photographs, and other material from 1935 to 1945 concerning Matanuska Colony.

Miller, Orlando W. *The Frontier in Alaska and the Matanuska Colony*. New Haven, Connecticut: Yale University Press, 1975.

National Park Service. *Alaska's Matanuska Colony*. By Darrell Lewis Edited by Janet Clemens. 2020.

Paige, Sidney, and Knopf, Adolph, "Reconnaissance in the Matanuska and Talkeetna basins, with notes on the placers of the adjacent region," in *Report of progress of investigations of mineral resources of Alaska in 1906, U.S. Geological Survey Bulletin 314*, U.S. Geological Survey, 1907.

Rollins, Alden M. *Census, Alaska*. Anchorage, Alaska: University of Alaska, Anchorage, 1978.

Smith, Arthur. *The Story of the Matanuska Valley Settlement*. Masters thesis, Occidental College: 1939.

United States Department of Agriculture, Soil Conservation Service. *Physical Land Conditions in the Matanuska Valley, Physical Land Survey No. 41*. Washington, D.C.: GPO, 1945.

United States Department of Commerce, Weather Bureau. *The Climate of the Matanuska Valley: Paper No. 27*. Prepared by Robert F. Dale, Washington, D.C.: GPO, 1956.

United States Department of the Interior. *Possibilities of New Land Development in the Matanuska-Susitna Borough, Alaska*. Washington, D.C.: GPO, 1970.

Bibliography

United States Department of the Interior, Bureau of Land Management. *Alaskan Group Settlement, the Matanuska Valley Colony.* Prepared by Kirk H. Stone. Washington, D.C.: GPO, 1950.

United States Geological Survey. *Preliminary Report on the Geology and Ground-Water Sources of the Matanuska Valley Agricultural Area, Alaska.* Prepared by Frank W. Trainer. 1963

United States Geological Survey. *Geology and Ore Deposits of the Willow Creek Mining District, Alaska,* Richard G. Ray, 1904.

Unites States Geological Survey. *Gold Mining in the Willow Creek District.* Stephen R. Capps, 1917.

United States Geological Survey. *Reconnaissance in the Matanuska and Talkeetna Basins, with Notes on the Placers of the Adjacent Region.* Sidney Paige and Adolph Knopf. 1907.

University of Alaska, Institute of Agricultural Science. *Climatic Characteristics of Selected Alaska Locations.* Prepared by C. E. Watson, C. I. Branton, J. E. Newman. Fairbanks, Alaska: 1971.

Alaska Agricultural Experiment Station. *Matanuska Valley Memoir.* Prepared by Hugh A. Johnson and Keith L. Stanton. Palmer, Alaska: 1955.

Williams, David. Papers. Louisiana State University, Baton Rouge, Louisiana, microfilm copy in Archives and Special Collections, Consortium Library, University of Alaska Anchorage. Copies of personal and business correspondence, reports, photographs. Williams was supervising architect for FERA.

www.ingramcontent.com/pod-product-compliance
Lightning Source LLC
Chambersburg PA
CBHW072151070526
44585CB00015B/1091